NATURE SPIRITS
WANT TO TALK WITH US

NATURE SPIRITS
WANT TO TALK WITH US

Listening Journeys in the
More-Than-Human World

COLEEN DOUGLAS

PLATYPUS
PUBLISHING

Copyright © 2024 by Coleen Douglas

All rights reserved. This book or any portion thereof may not be reproduced or used in any manner whatsoever without the express written permission of the author except for the use of brief quotations in a book review.

Published by Platypus Publishing LLC
Contact the author at www.coleendouglas.com

ISBN (paperback): 978-1-962133-87-6

Cover Photograph by Coleen Douglas
Interior photographs by Coleen Douglas, Dawni Pappas, Ellen Farmer
Copyediting by Melody Culver, Writers Services
Book Design and Production by Constellation Book Services

Printed in the United States of America

Grateful acknowledgment is made for permission to reprint from the following:

Spangler, David; Jeremy Berg, art. Card Decks of the Sidhe Manual. Douglas, MI:Lorian Press, 2011. https://lorian.org/bookstore-sidhe

Spangler, David. Engaging with the Sidhe: Conversations Continued, Douglas, MI: Lorian Press. https://lorian.org/bookstore-sidhe

Charters, Daphne. Forty Years with the Fairies: The Collected Fairy Manuscripts of Daphne Charters, Vol.1. Compiled by Michael Pilarski. West Virginia: RJ Stewart Books. https://rjstewart.net/forty-years.htm

To our collaborative spirits

CONTENTS

Acknowledgments	ix
Preface	xi
Invitation	1

Part One: My Learning Journey — 3
- Chapter 1: Science-Trained Mind and Nature-Loving Heart — 5
- Chapter 2: Tune In to Subtle Realms of Redwoods — 15
- Chapter 3: What Are Nature Spirits? — 22
- Chapter 4: Why Do We Need to Hear From Nature Spirits Now? — 33
- Chapter 5: Energy Healing And Receiving Messages In The Forest — 38
- Chapter 6: Sacred Forest Song — 49

Part Two: The 8 Timeless Principles — 53
- Chapter 7: Principle 1: Release Judgment — 55
- Chapter 8: Principle 2: Respect — 59
- Chapter 9: Principle 3: Love — 63
- Chapter 10: Principle 4: Ask — 67
- Chapter 11: Principle 5: Listen — 71
- Chapter 12: Principle 6: Trust Your Intuition — 84
- Chapter 13: Principle 7: Imagine — 89
- Chapter 14: Principle 8: Cultivate Relationship — 97

Part Three: Develop Your Ability to Communicate — 105
- Chapter 15: Going Deeper — 107
- Chapter 16: Ancestors — 120
- Chapter 17: Co-Creating A New Earth with Nature — 139

Conclusion	*161*
Practices By Chapter	*163*
Notes By Chapter	*167*
Further Learning Resources	*176*
Meet The Interviewees	*184*
About The Author	*191*

ACKNOWLEDGMENTS

Gratitude to the many teachers in my life. For my journey into relationship with nature spirits, I thank these beautiful ones: Dawni Pappas, Camilla Blossom, Barbara Thomas, Kami McBride, and the Redwood Forest. Thank you, nature spirits, for your willingness and connection—life is full of wonder and joy in your presence. Thank you to all the people who will come into my life as I continue to learn about nature spirit communication.

I so appreciate the thirteen people I interviewed about their experiences communicating with nature spirits. Their openness gives us insights into the wide variety of ways this communication occurs. (Read more about them in Meet The Interviewees.) Thank you to the Fairy and Human Relations Congress and Gaian Congress for bringing together people and nature spirits to learn how we can work together for healing and regenerating the earth.

With great gratitude and love I thank my writing group for their incredible creativity and encouragement: Chopsy Gutowski, Ellen Farmer and Sandia Belgrade. Many of the poems in this book were written during our time together, and each of these women offered essential advice about my writing. Although we began writing together twenty years ago, that stopped when one of us moved to a different state. Our writing group got together again on Zoom starting in April 2020 during the coronavirus pandemic, and we have been meeting together weekly since then.

Originally, I planned to write a magazine article about nature spirit communication, but that didn't pan out. I would not have attempted to write this book without Matt Rudnitsky's Punchy Book Accelerator course (although it took me much longer than I expected to complete it). I learned so much in a way I could understand about how to write a short nonfiction book and how to self-publish. I thank my "accountability buddy" in this course, Patricia Pfost, who met with me regularly and gave wonderful encouragement and advice. I also received generous guidance from Edveeje Fairchild, Donna DeBenedetto Barwald, Alanna Sharpe Bell and many others on the private course Facebook group, with 3,000 members who are willing to support each other.

Thank you Jenny Kirk, Deborah Helms, and Barbara Crum for reading the entire draft manuscript and giving indispensable feedback from the reader's perspective that made the book flow. I appreciate the many friends who read a portion of the draft book and gave important comments: Andrea Boone, Christine Barrington, Cristina Case, Cyndy Crogan, Della Duncan, Irene Reti, Jean Mahoney, Jo-Ann Birch, Katherine Bell, Kendra Ward, Kim Calvert, Melissa Trail, Nancy Harper, Rebecca Gottlieb, Rhan Wilson, Sam Ambler, and Shinehah Bigham.

Thank you Melody Culver, for your astute copyediting and proofreading and for sharing your years of experience in book production. Many thanks to Christy Day and Constellation Book Services for bringing this book into physical form and beauty.

I am grateful to those who have taught me about Indigenous wisdom, including Gregg Castro, Stan Rushworth, Valentin Lopez, Kanyon Sayers-Roods, and members of the Cultural Competence

Circle of Ecosystem Restoration Communities California. Thank you, Stacey Wear and team from the Camp Fire Restoration Project, for including the Cultural Competence Circle final document on your website so anyone may have electronic access to it.

Thanks to all, seen and unseen, who have inspired me throughout my life to keep reaching for and living my full authentic expression.

Finally, I give abundant gratitude to my brilliant life partner Ellen Farmer, who continues to grow and change in her own ways, while loving and supporting me through my life changes. Your truth-telling with kindness is essential. We are beautiful together. I love you beyond words.

PREFACE

This book is for people who love nature and want to go deeper. It provides a step-by-step way to connect deeply with nature and open yourself to receive nature's messages.

My journey is a bridge connecting the science-trained mind and nature-loving heart. I am a nature lover. Going into nature gives me a feeling of wonder—from the magnificence of the ocean to the tiny intricacies of a flower. Trained in biology, I love to observe the interconnectedness of plants and animals in their ecosystems. When I began going into the forest with a close friend whose experience was in the mystical realms, my heart and mind started opening to nature spirits and the possibilities of communication between us.

My science-trained mind had such a hard time opening to the unseen world of nature spirits. That's understandable, because in today's world, especially in the U.S. where I live, we are saturated with distrust of anything different than the individualism, economic imperatives, need for proof, and power dynamics of the dominant culture. Modern society is only beginning to understand what our ancient tribal ancestors knew and practiced—we people are intimately connected with nature and spirit. Those of us moving in the direction of interconnectedness have a sense of this ancient truth.

As you read about my learning journey and the 8 Timeless Principles for communicating with nature spirits, you will find experiential ways to put this information into practice in your own life. I've been on this journey for a decade and have worked with several teachers to go deeper. For this book, I interviewed thirteen experienced nature communicators whose wisdom is shared within. I include further resources to develop your ability to communicate, including Notes by Chapter at the back with references for specific topics, quotes, and people mentioned in each chapter. You will also find Practices by Chapter at the back, listing the nature spirit communication practices found in each chapter.

We all depend on nature for life. I am learning how to listen to nature spirits to understand what I can do to best support what nature needs. **Nature spirits want to talk with people now because we have a lot to do together to restore the natural world devastated by our human actions.** For those people who have found their place in nature and are already contributing to solutions, thank you.

What if we all consciously allied ourselves with nature?

As we learn to listen to nature and her spirits, **we can uncover key messages about regenerating life—our own personal lives, and the life of the natural world.** Learning to listen to the more-than-human world is a journey into self-expansion and restoration. Learning to listen to the more-than-human world is also a continual source of connection and joy.

INVITATION

Once when I was called to lie on the forest floor, redwood tree sent a message to me through her roots and her network. Hands pressed against bare earth, I felt a tingling in my fingers, and this word appeared—*Flow*.

The tree is inviting us all
To open and allow ourselves to flow
Live life in the flow
Let love wash over us and seep inside

We are like the forest creek gurgling in our ears
Gather all the water drops
To flow together around barriers
Down to the lowest point
Into the sea
Into the arms of the mother of us all

Let go of resistance to the unknown
And just flow
Life is what we live right now
Moment to moment
Set your intention on love for all beings
And flow

I invite you to come into nature with me in a different way. Let nature's beauty fill you with awe and wonder. Let nature spirits talk with you in a way you can understand. There's so much possibility when we take guidance from nature to live our lives with new eyes, open hearts and curiosity.

What could we learn if we could talk directly with nature spirits?

This book is a guide for delving deeper. We'll explore together the 8 Timeless Principles for communicating with nature spirits. Let's go!

PART ONE
MY LEARNING JOURNEY

CHAPTER 1

SCIENCE-TRAINED MIND AND NATURE-LOVING HEART

My story is a bridge connecting a science-trained mind and nature-loving heart. At this point in my life, I can cross this bridge in either direction. There are gifts on both sides.

Back then...

I grew up in the suburbs of Los Angeles, and as a child I loved to climb trees. In our yard I would climb up and sit in their branches, with leaves all around me—my own space, away from people and expectations. The trees held me. I felt safe and fully accepted. I could look out through the leaves for a different view of my world. When it was time to come back inside, my mom would call us three kids, hands cupped around her mouth: "Bobby, Billy," then facing up to the treetops, "Coleen." She knew.

In 1971, I left the hustle and smog of Los Angeles to go to college at the new University of California campus in Santa Cruz, where the buildings were constructed at the edges of a redwood forest. I knew this was my place. I focused on the sciences and walked through the redwood forest to classes.

In my first year, in a class on ethology (animal behavior), I found myself in the Science Library going through the stacks of journals, writing a paper on the symbiotic relationship of sea anemones and clownfish. (This was thirty years before the film *Finding Nemo*.) In the coral reef, the anemone defends itself with poison nematocyst stingers at the tip of each tentacle. However, the mucus layer on clownfish skin protects it from the anemone's stingers. The anemone lets the orange clownfish shelter among its tentacles, protecting it from larger predators. In turn, the clownfish cleans parasites from the anemone, provides food (nitrogen from its waste) for growth and regeneration, and helps circulate water around the anemone. Symbiosis. Collaborative relationships. This was the kind of relationship that grabbed my attention.

Walking in the redwood forest, I would stop, and sit, and look. I moved the tangled sticks and branches so I could be close, and even inside, these beautiful trees. I heard birds but rarely saw them. I watched flowers emerge through redwood duff in spring, and mushrooms in winter. I would lean my back against the massive trunk and look up, look out in quiet wonder, feeling this amazing place.

I studied marine biology. I also took pre-med courses because my parents thought I should go to medical school. I liked learning and practicing the scientific method: thinking of a creative hypothesis, testing it out, and being able to prove or disprove my initial hypothesis.

Late in my college career, I finally took physics. I didn't like the course because it just didn't seem to have anything to do with living things. I was absolutely flabbergasted to learn about the "observer effect": when an electron is passing through an opening, sometimes it appears as a wave and sometimes as a particle. *How the observer looks at it determines how it shows up.* Learning that the observer seems to have a big impact on an experiment, I realized that there is so much more beyond the scientific method. This was really my first knowing that the "mystique of science" was not the whole truth.

After graduation, I worked in a marine science company for a few years. I was excited about putting what I learned in college to work in the "real world." But the real world of commercial marine science turned out to be quite different than college science. The scuba diver biologists set up transect lines to measure the animal and plant life in different places that could be affected by human impacts. They would then scrape off the surface of the rocks in square samples and bring back the scrapings to identify what was there. My job was to identify the dead beings sitting in alcohol or formaldehyde, back at the lab. I would sit at a microscope, breathing the sickening fumes, identifying plants and animals that had previously been alive.

After several months of this soul-deadening work, I asked if there was something else I could do. I got the opportunity to go out in a small boat as part of the crew taking water samples at different depths. That sounded exciting, to be out there on the ocean! When we stopped at different locations for water sampling, I discovered that I got seasick when the boat was anchored. In college, my fieldwork had been in the intertidal zone—at the ocean shore, not on a boat. I couldn't wait to get back to land. Ultimately, I ended up being the reports coordinator for the company, helping to clearly document the marine science work that had been done.

The unintended consequence of the marine science company was that I became a singer in my first band there. Everyone in the band worked there, and the company let us practice in their conference room one night a week. I am still friends with most of those band members today. This music connection was an absolutely critical aspect of my life. For me, it turns out that music is one path to the subtle realms of nature.

While working in marine biology, in the late 1970s I became one of the volunteer "founding mothers" of the Santa Cruz Women's Health Collective, providing health care and health education for women. It was the height of the Women's Movement, and we were a very political group of women challenging our local male doctors' stranglehold on health care. "Health care for people not for profit" was one of our strongly held beliefs.

Because of my parents' desire for me to be a doctor, I applied and got accepted into one medical school. After visiting the campus and thinking long and hard about it, I decided not to go. I realized this was my parents' dream for me, not my own dream. I loved the alternative health care path I was on with the Women's Health Collective and did not want to abandon that for a path that recognized only western medicine. I knew that healing was so much broader than that. This was the best decision I ever made about my life's path.

...And now

Following my own dreams, in the past five years my life's path has taken me into the world of ecosystem restoration and a greater understanding of Indigenous wisdom. Weaving my life experiences, I see that these two strands have opened me to deeper relationships with nature and her spirits. Ecosystem restoration and Indigenous wisdom expanded me in both directions: healing the land while feeling a spiritual connection to all life, and nurturing community while opening our hearts, minds and spirits to each other.

ECOSYSTEM RESTORATION

In 2015, I was inspired by a documentary film by John D. Liu called *Green Gold*, about the massive project to restore the Loess Plateau in China. The restored area, the size of Maryland, had been the most fertile land imaginable, but after decades of uncontrolled grazing by goats, the plant life had been decimated and all that was left was barren ground. It took ten years, but you can see the change happening before your eyes in this film. It took thousands of people to restore these ten million acres, to halt erosion, restore the land, and return water to the land. The people became able to support themselves again. And the world could see that large-scale ecosystem restoration was not only possible but also proven. This gave me tremendous hope for the future of nature's bounty and what people can do together.

Filmmaker John D. Liu started a movement to develop Ecosystem Restoration Communities worldwide, with the first one in Spain. In March 2019, I was part of the California Ecosystem Restoration Council, the first organizing gathering to develop Ecosystem Restoration Communities in California. One hundred people came from all over the state to Paicines Ranch, a place dedicated to developing regenerative practices in agriculture. John Liu was present, and among others, provided inspiration, films and photographs of how large-scale restoration could be done. During that weekend, the first California Ecosystem Restoration Communities project was organized led by Matthew Trumm, a permaculturist who lives near Paradise, California. The previous November 2018, the town of Paradise had burned to the ground in a massive wildfire, the Camp Fire, killing at least eighty-five people and destroying 19,000 homes and businesses, along with trees and the land.

The Camp Fire Restoration Project's first work weekend took place in April 2019, only one month after the original council meeting and six months after the fire. It was an amazing weekend of learning, restoration, and community building. I helped organize the Healing Tent, where we provided herbal remedy teas, massage, energy healing, resources, and deep listening for the people traumatized by the fires and the people doing restoration work that weekend. My singing partner Nita Hertel and I led songs for the community to bring people together and lift spirits. People from all parts of California were involved, and some were inspired to begin similar Ecosystem Restoration Communities in other parts of the state.

Under the leadership of Janeva Sorenson, Stacey Wear, and their team since 2020, the Camp Fire Restoration Project has grown into an organized 501(c)3 nonprofit organization that is expanding beyond disaster response and becoming a long-term resource for ecosystem resiliency. They focus on action-based projects for nurturing soil health, trees and seeds, water and community. The Camp Fire Restoration Project hosts Paradise Community Compost, a practical and educational program working to promote soil health, stewardship, and organics recycling. They also provide support around Indigenous Traditional Ecological Knowledge (TEK) in partnership with Ali Meders-Knight, a local Mechoopda tribal member and Master Traditional Ecological Practitioner.

INDIGENOUS WISDOM

In 2020 we were confined to our homes during the pandemic, and community gatherings were stopped. As online Zoom gatherings began, Ellen Farmer and Anna Hope organized an online Cultural Competence Circle for California Ecosystem Restoration Communities. Thinking ahead to future in-person community efforts, they knew that this work on land in California is always happening on unceded land of Indigenous people. "Unceded land" is a short phrase but its meaning is complex—traditional lands taken illegally from Indigenous people by violence, unfulfilled treaties, cultural erasure, enslavement, and genocide. How might it be possible for we non-native people to learn to partner with local Indigenous people, to respect and honor their lives, traditional lands, and traditional ecological knowledge? We knew we needed to expand our knowledge and practices.

Our group of twenty-four people met online every two weeks from April 2020 through February 2021. We were blessed to have two California Indigenous guides offer to be part of our group: Gregg Castro, an elder and wisdom keeper (t'rowt'raahl Salinan / rumsien-ramaytush Ohlone) and Kanyon Sayers-Roods, artist, consultant, and daughter of Ann-Marie Sayers (Costanoan Ohlone-Mutsun & Chumash). This online group with participants from Los Angeles to Mount Shasta was one of the hidden blessings of the beginning of the pandemic. Over the year we opened our hearts, minds, and spirits to each other. We began to develop trust, the most essential key to building relationships.

Based on extensive notes from our meetings, we created the document "Land Acknowledgement and Relationship-Building Protocols with Indigenous Peoples at Ecosystem Restoration Communities California." It was reviewed by a number of Indigenous people, as we wanted to share it broadly. This summary sentence from the introduction, is written by Gregg Castro:

> It is an offering of our collective thoughts and experiences in our journeys through life and how we might best forge deep and lasting bonds with each other and the land we are healing.

This process influenced participants in California Ecosystem Restoration Communities, as well as movement founder John Liu, to begin or continue the process of partnering with Indigenous people, to learn with and from them. Developing these relationships takes time, often years. I know of large organizations who offer paid consultancies to Indigenous people for their expertise, time, and travel expenses. Everyone wants to be treated with respect for their contributions.

For me, this learning process opened my ears and heart deeply to Indigenous wisdom and the true history of Indigenous peoples of California. Their land was taken from them without compensation. They were enslaved by the Spanish California Mission system, and 90–98% of Indigenous people in their regions died as a result. I learned about the state-sponsored genocide of native people in California, where the state offered "$5 per scalp" to remove native people from their land during the gold rush. As a result, "pioneer" immigrants used the resources of Indigenous lands to build towns and create wealth. And now, the descendants and new settlers of California are seeing the results of greed and ignoring native wisdom: wildfires, desertification, the poisoning of soil, air, and water.

Once I learned the truth, I could not unlearn it. Mission Santa Cruz, in my town, was known as the most brutal of the California missions. Now I talk to friends and colleagues about the truth of Indigenous genocide in California and continue to study and learn about Indigenous wisdom. I support the descendants of our local tribes—the Amah Mutsun Tribal Band—by volunteering in their projects, giving monthly donations, continuing to learn of Indigenous traditions, and by unlearning my old ways. I encourage everyone to learn about the Indigenous history where they live and work.

Now is the time for understanding the wisdom and practices that Indigenous people of California have used through time immemorial to create and maintain a paradise here on earth. They are following Creator's instructions to care for the land and all the life within it—knowing that all of nature is alive and our relatives, kin.

A portion of the introduction to "Land Acknowledgement and Relationship-Building Protocols with Indigenous Peoples at Ecosystem Restoration Communities California" is a good example of the Indigenous approach:

Introduction by Gregg Castro

Shaamo'sh Ake'sh/ Mishshix tuux / Hersha Tuhe— Greetings, 'Relatives'. As we are all 'two footers' (as my Elder-Mentor named us 'human beings'), we are related through that common humanity. We are then connected through the common landscape that we walk, breathe, think, love and pray on. This connection is not a static strap, it is a dynamic, living umbilical cord that binds us together in a weaved basket of life. One that is constantly inhaling and exhaling, writhing and changing. If one opens not just their eyes, but their minds, hearts and spirits to this seething sea of life around us, we learn that we are an inexorable part of it all, inseparable but not indomitable in its hierarchy. As my Elder-Mentor said often, most of the origin narratives of California Indigenous Peoples say that we came last in the creation of the world, and this gift—that we need to accept humbly and gratefully—is to be taken care of in deep respect and humility as all gifts should be.

Now there are many ways to learn from Indigenous people: webinars, websites, books, audiobooks, TEK workshops (Traditional Ecological Knowledge), podcasts, and more. A recent online search for "Indigenous Wisdom" generated 44 million results. Please go into your process by learning directly from Indigenous people. You'll find recommendations to get you started at the end of this book, in Further Learning Resources. For now, I want to make sure you know about this great book/audiobook: *We Are the Middle of Forever—Indigenous Voices from Turtle Island on the Changing Earth*, edited by Dahr Jamail and Stan Rushworth.

The book draws on interviews with people from different North American Indigenous cultures and communities, generations, and geographic regions who share their knowledge and experience, their questions, their observations, and their dreams of maintaining the best relationships possible to all of life.

READER'S QUESTION FOR REFLECTION

How has your life's journey shifted by being inspired through your learning and actions?

CHAPTER 2

TUNE IN TO SUBTLE REALMS OF REDWOODS

My journey to learn to communicate with nature spirits began on walks in the redwood forest with my friend, Dawni Pappas, where we live in Santa Cruz, California. I was used to walking in the forest, looking closely at the plants, learning their names and wondering about how they all lived together. I just loved being completely surrounded by nature, feeling the wonder of walking among the big redwoods and their forest mates. Dawni had many years of experience in the mystical realms, and over time she invited me to connect with the forest in new ways. Dawni would tell me:

> Let's stop here at this circle of redwoods. Let's ask permission from them to go inside their circle, their fairy ring. Let's use our senses to hear yes or no. It's good to be able to tell the difference. Feel a 'yes'? Good. Step inside.
> Now, let's say hello to them, and send them our love. Feel your heart opening, and the love energy connecting us. How does that feel?
> Now, let's "shake hands" with them by gently stroking their leaves. They love it when we do that. What happens inside you when you share this sensual physical touch?

Let's give one of them a hug now. What do you sense when you put your arms around their trunk, and rest yourself against their living wood?

Well, this was new to me! I was willing to try but felt a bit shy about it. Over time, this kind of sensory connection with the forest became absolutely natural. I loved being with the trees and began to feel their loving energy in return.

Later, Dawni talked about how we could ask for messages from the redwood trees. "Really?" I said. She had been asking for and receiving messages from nature for a long time. We lay down on the forest floor, breathed in and out to come into a state of relaxation, and connected with the trees using all our senses, one at a time. Then, listening with all my senses, I asked the redwoods, "Do you have a message for me this day?" The word "Flow" came into my mind. When our meditative time ended, we talked about what happened. I wondered if I thought up that word myself or truly received it from the trees. But why would I make up that word? As I considered it further, I felt that it was a message for me to flow with whatever was happening there in the forest. *Don't worry about it, don't try to control it, just go with what's happening. It's okay. It's all a learning process.* This is not the only time I've received the message "Flow" from the forest. I think I need that reminder the more I open to new experiences in nature, and the further I go in my learning journey.

One day, Dawni heard the forest call out to us: "Bring people back to nature, and nature back to people." Over the next few months, we developed an airbnb experience called Tune In to Subtle Realms of Redwoods. Now we guide people from all over the world into the redwood forest to connect deeply with nature. This is how we describe our experience:

Is the redwood forest calling you? Would you like to know it in a new way? In this guided experience you will make a personal connection with the subtle realms of the redwood forest. Going into the redwood forest together, we will walk, pause, energize all our senses, and attune with the subtle realms of nature. The gift of music will set the tone, and our loving presence will open the way. We will call on the elements earth, air, fire and water to bring us to a new understanding of the aliveness and interconnectedness of the natural world.

We will spend a few minutes writing about our experience to help you to recall how to connect with the subtle realms back home. As we complete our time in the forest together, you can choose to stay longer on your own, fully energized with a new vision of our interconnectedness.

We developed an airbnb experience called Tune In to Subtle Realms of Redwoods. We now guide people from all over the world.

We've been doing this since 2018, and we love it! It feels right for us to do what nature called us to do.

At the beginning, I would offer scientific explanations and Dawni would offer her understandings of the mystical. The more I went to the forest in a loving and open way, the more I became integrated with all the life in the forest. I started singing us into the forest, providing the magic of song floating above us as we walked in more deeply. I know how music goes directly into people's hearts while bypassing their brains, helping to bring us to another level of connection. It was exciting to become attuned to the redwood trees, beings who have been around for, in some cases, a thousand years. Dawni regularly received messages from them, and I was starting to as well. Sometimes it was a word or phrase, sometimes a vision. We now invite our guests to ask for a message from the redwoods, listening with all their senses.

I wrote this poem to give a sense of the kinds of things we do as we Tune In to Subtle Realms of Redwoods:

THE ELEMENT OF AIR

Come into the redwood forest with us
Put your phone on airplane mode for these few hours
And be fully here
Where the tallest trees in the world live
In community with all beings, seen and unseen

Let's walk a little ways together
Here we are at the overlook
Where we can see the tops of the redwoods rooted down the hill
Coast redwoods, Sequoia sempervirens
They can be 300 feet tall
That's 100 yards, the length of a football field

TUNE IN TO SUBTLE REALMS OF REDWOODS

Imagine one lying on its side covering the entire field

Here in the forest today
We will be connecting with all our senses
We'll focus on hearing, touch, smell, and taste, as well as sight
We'll bring forth the elements into our consciousness
Earth, air, water, and fire
And the fifth element—love
Love, the power that unifies
Love, the connecting impulse
For we are intimately connected with all the beings in the forest

Let's start with the element of air

Take a deep breath of this forest air
In breath—out breath
Can you smell the freshness?

As we look out over the redwood forest,
let's take another deep breath of forest air
In breath—out breath
How does the air taste here inside the forest?

And one more deep breath
In breath—out breath
How does this air feel within your body?

With each breath, these redwoods, right here around us
Are making the oxygen we are breathing in
Yes, all the green plants in the world are as well
But...these redwoods standing tall around us, right here
It takes only one minute—60 seconds—for their oxygen
To circulate through our entire body
From our lungs to our bloodstream

Flowing to all parts of our body to feed our cells
Then flowing back to our lungs
To release carbon dioxide into the air—our "waste product"

Carbon dioxide is what trees need to make their food
Carbon dioxide plus water
With power from sunlight
Turns into sugars that feed the leaves, stems and trunk of the tree
Through the magic of photosynthesis
Then 20% of those sugars are released out through their roots
Feeding the fungi and microbes below ground
Who create a massive underground internet
Connecting roots to roots, plants to plants, all over the forest

And the "waste product" of photosynthesis?
Oxygen
Released into the air to allow life
For us humans and all the animals

Now, let's take another deep breath of forest air
Fully aware of our connection to these trees
In breath—out breath
We are intimately connected with these redwood trees
At the molecular level
Every time we breathe

This is the element of air

For me, the redwood forest was the perfect place to begin learning about nature spirits. I already had a deep connection to the forest, and to the variety of plants and animals that live together in this interconnected beauty. The spirits of trees who send me messages are nature spirits. They are invisible. But the trees are not;

I can see, touch, smell, taste and hear them. It was just a matter of time before I could open myself more and more to the diversity of nature spirits, unseen but all around us, all the time.

READER'S QUESTION FOR REFLECTION

Where in nature do you feel a special connection?
How do you feel when you're there?

CHAPTER 3

WHAT ARE NATURE SPIRITS?

Everything in nature has a spirit, just as everything in nature is alive and interconnected. All my life experience has led me to this conclusion.

Although I grew up going to church, by the time I got to high school I had rejected the organized-religion form of spirituality. As I spent more and more time in nature, I felt deep in my soul that god is "the interconnectedness of all life." Throughout my life, I've learned about many different spiritual beliefs, experienced a variety of spiritual practices, and developed personal spiritual practices that fit for me. Now I experience the deepest spiritual connection in nature.

Being alive in the United States for decades, I am steeped in modern culture. So much focus on scientific materialism, reducing our human existence to solely physical existence—no spirit, no spiritual beings. At the same time, many prominent organized religions are monotheistic—only one god, their God, denying the existence of all others. I wonder what my long-ago tribal ancestors knew about nature spirits?

I love how John Trudell (Santee Sioux) describes the history of humanity:

All human beings are descendants of tribal people who were spiritually alive, intimately in love with the natural world, children of Mother Earth. When we were tribal people, we knew who we were, we knew where we were, and we knew our purpose. This sacred perception of reality remains alive and well in our genetic memory. We carry it inside of us, usually in a dusty box in the mind's attic, but it is accessible.

We all have the opportunity to shake off the "dusty box in our mind's attic" and access the ancestral sacred perception of reality. As I continue to open myself to be intimately in love with the natural world, my spirit is overjoyed. And so are the nature spirits.

THE DIVERSITY OF NATURE SPIRITS

We've heard of fairies and gnomes, mostly as characters in fairy tales or movies. They are much more than characters in stories; they are nature spirits. I am learning a lot about the huge diversity of nature spirits. Everything in nature has a spirit. For example:

The spirits of trees, flowers, moss, fungi, seaweed, birds, spiders, bears, whales—all plants and animals—are nature spirits.

The spirits of the elements—water, fire, air, earth—are diverse and have many names and appearances, such as water deva, undine, Yemaya, mermaid; fire dragon, sun; air dragon, sylph, fairy, wind; crystal, gnome, Sidhe, stone.

The spirits of ancestors—my own ancestors, Indigenous ancestors, star ancestors—are nature spirits.

People categorize some of these beings as different types of spirits. For this book, I call them all nature spirits. People also use words with different language origins to describe nature spirits as a group or individually, such as fae, fay, fey, fair folk, faerie, faery, fairie, fairy. If you'd like to learn more about this diverse realm, I highly recommend Camilla Blossom's 2022 book, *Sacred Spirits of Gaia: Co-Creating a New Earth with Fairies, Elementals, Ancestors and Spirits of Land*. This powerful collection of information about many nature spirits includes ways to co-create with the seen and unseen realms of nature through intuitive techniques, land rituals, land alchemy practices, tools and wisdom teachings.

MY GROWING RELATIONSHIP WITH THE SIDHE

My friend Dawni introduced me to the Sidhe (pronounced *shee*) seven years ago, by sharing her Sidhe oracle cards with me. I had occasionally used tarot cards as an oracle to see what guidance they might have but had never heard of the Sidhe. The Sidhe oracle cards are called *Card Decks of the Sidhe* by David Spangler, with art by Jeremy Berg. Both the words in the book and the beautiful artwork for each card (a collaboration between David, Jeremy and their Sidhe partner Mariel) have become meaningful and useful in my life.

Sidhe is an ancient name for the faerie races of Ireland. They descended from the people of Danu, or the Tuatha Dé Danann. They are considered ancestors, gods, goddesses, and spirits of nature as they were the pre-Celtic inhabitants of Ireland. They retreated into invisibility in underground mounds. Sidhe means "people of the mounds." There are many names for the Sidhe, including "people of peace" and "the shining ones." Connecting with the Sidhe centers you in the early traditions of Ireland and Scotland.

As my father's ancestors come from Scotland, the Douglas clan, I feel a particular connection to the Sidhe. I bring out the Sidhe cards often, seeking guidance or inspiration for my writing, work with nature, or daily life. I often choose one card that is calling to me at that moment, and the art and words of this oracle speak to me deeply. I have also spent several days on retreat, connecting with the Sidhe through complex layouts of the Sidhe cards.

I wrote the following poem about my relationship with the Sidhe oracle cards, from the standpoint of the oracle cards themselves:

ORACLE CARD DECK OF THE SIDHE SPEAKS TO ME

You've been calling on me for guidance
These past 7 years
When you need a message for your soul
When you need inspiration
When you want to know what to do next
When other methods of communication aren't coming through
Just skip the technology
I am your landline
Connecting spirit and body
Direct and personal

You find me easily
Open the rough fabric sleeve of protection
And pull out my oracle cards
You finger them lovingly
Knowing the wisdom of the Sidhe lies here
Ready and willing
You give thanks for my oracle

Holding me in your hands
You close your eyes and connect
The visitation begins
You ask your question
Its energy flows through your hands into me
There's imperceptible movement within me
Among each of my cards
Then you spread out my cards face down in a rainbow arc
Each one visible in some way

You focus on the energy in your hand
As you sweep your hand slowly over each card
Energy and a finger twitch show your card for today
You flow over the whole deck to see if there's only one

Like today, you asked for a message for guidance for this week
Busy with new challenges unplanned
With people to lead
With compassion and love to express
With sound equipment and shovels
With trees surrounding all
My *Stone of Identity* card and *Palace* card are both needed
For you to see and feel the fullness of this time

Stone of Identity card addresses
Your personhood, the partnership of soul and body
Shown as falcon and standing stone
Spirit and earth
This unique amalgam of you as human
I show you how broad and expansive you are
So that you may bring the wholeness of you to this week
You have everything you need

Palace card shows
Your dedication to harmony and beauty
You are a wise teacher
You are a leader
You are an ally
You are able to take on the responsibility

All is well
I feel you relaxing here into the truth
The Sidhe are here for you
As always
Through me, the oracle

IF WE CANNOT SEE THEM, HOW CAN WE COMMUNICATE WITH NATURE SPIRITS?

Nature spirits are not visible to the human eye because they vibrate at a higher frequency than we can detect with our eyes.

From science, we know that all molecules vibrate. Every molecule in our human body is vibrating at a certain frequency. Although a table looks completely solid and is not moving around by itself, it is made of molecules that are vibrating; they are just vibrating at a lower frequency than we are. Nature spirits are vibrating at a higher frequency than we humans are, so we cannot see them with our physical eyes.

If we cannot see them, how can we communicate with nature spirits?

I have found that it's about being in a state of love, respect, and openness to nature spirits. And it's about being open to receive communication in any form: words, visions, feelings, vibrations, colors, sounds, or any other way.

FROM DOUBTS TO OPENNESS

As you've read in previous chapters, my learning journey has taken me through disbelief and doubts to the openness needed to know nature spirits more profoundly.

One thing that has helped me is to talk with people who have been connecting with nature spirits for many years. I interviewed a number of people for this book who are experienced nature spirit communicators. You will find their experiences and ideas throughout this book, and you can find out more about each interviewee in Meet The Interviewees.

Here are a few of their insights on the question "What are nature spirits?"

Lindsey Swope

There is intelligence in nature. Nature is not a machine. Science has been showing that nature has intelligence, that it has an understanding…like trees reach for light, they reach for water, they share resources. So there's this intelligence. And nature spirits are really just how we interact with that intelligence.

Barbara Thomas

Nature spirits are nonphysical conscious beings within every aspect of nature.

Dawni Pappas

Nature spirits are like music. Music is invisible. It comes from seemingly nowhere. Music is an energy, a frequency, a vibration. You're considered gifted if you can hear music and make music, and then be able to write it down and do it again. The invisible made visible. When it is made visible, it takes on a life of its own. It can catapult me into a new dimension.

I'd also like to introduce you to Daphne Charters, an ancestor nature spirit communicator who has influenced many people. She lived in England from 1910 to 1991 and was known for communicating with a group of nature spirits—fairies. She had conversations with them regularly and documented everything she learned from them. In 1956, she wrote a book called *A True Fairy Tale* documenting her key experiences (reprinted in the 2008 book, *Forty Years with the Fairies: The Collected Fairy Manuscripts of Daphne Charters*, Vol.1).

Lord Hugh Dowding, a well-known high-ranking officer and Chief Air Marshal for Britain during WWII, believed in fairies and wrote the introduction to Daphne's book:

> I fear that this book is years ahead of its time. People are so hard-headed! Fairies are seen today in thousands of cases, but mainly by very simple people, as for instance, Celtic peasants and children of all nationalities. Now and then they are seen by psychic adults. And yet, to say of a person, "He believes in fairies" almost amounts to saying, "He isn't right in the head." Personally, I believe that we shall all eventually come not only to accept the existence of fairies as a fact, but also to recognize the essential work which they do for humanity, and to accept the obligation therein implicit to afford them our loving cooperation...

I'd like to share brief excerpts from Daphne Charters' 1956 book that are helpful to me regarding the question: "If we cannot see them, how can we communicate with nature spirits?"

> I am not clairvoyant in the usual sense. I can, however, "see" a little with my mind. If I say to you, "Last week I watched the sun setting between two ranges of snow-capped mountains and the lake lying between looked as though it were on fire," unless you are very unimaginative, I feel sure that some kind of picture will take form in your mind. In other words, you will be seeing with your "mind's eye" as opposed to your physical or psychic eye. This is how I "see."
>
> *Vibrations*
>
> This ability to see is not due entirely to the mind's eye but to vibrations…I am sure that some of you must have entered an apparently empty room and yet have known that you were not alone. This is often called the sixth sense but actually you are receiving the invisible vibrations of the other person present…

Mediumship

I expect that most of you have a radio and you will know that the waves of sound are captured by your set so that you can hear music and talks coming from many miles away. These waves or vibrations are in your room all the time, but you cannot hear them unless you tune into them. In other words, your radio is the medium through which your ears are enabled to receive the sounds which are already present.

Being a medium merely means that a person is able to receive sights and sounds not visible or audible to the normal eye and ear.

I have no doubt that many of you are convinced that fairies are imaginary beings which have been invented for the amusement of children and, in stating that I have conversations with them every day, I am in grave danger of being labeled an amiable lunatic.

However, it is not always wise to scoff at people who have access to experiences which, up to the present, you have not. I will take the question of hearing. Tests have been made by various groups of people and indeed you could carry out a similar experiment yourself, with the aid of several friends, if you are able to buy or borrow a super-sonic whistle. The pitch of this instrument varies from a tone which all with normal ears can hear, to so high a note that no human being can detect it. We will imagine that you are with eleven trusted friends and, to begin with, all can hear the whistle. Gradually the pitch of this instrument is raised until in ones and twos your friends drop out of the group, as the note becomes inaudible to them. You are the eighth to retire, leaving four who maintain that they can still hear the whistle. If you are a scoffer, you must now make up your mind whether:

(1) Your four friends are experiencing something beyond your own capacity.

(2) Their hearing is a figment of their imagination.

(3) They are just plain liars.

If you decide on (2) or (3), remember that the seven friends who dropped out before you are equally entitled to think the same thoughts of you…

I have stressed this point in order to dissuade you from laughing at children when they come and tell you of their fairy friends. Young people are often psychic up to the age of adolescence, so encourage them to share these playmates who, although unseen by you, are as real as you yourselves.

Inspired by Daphne Charters' work, the Fairy and Human Relations Congress has been taking place since 2000 in the Pacific Northwest area of the United States. The focus of this yearly in-person gathering is "Communication and Co-Creation with Nature Spirits, Devas and the Faery Realms." Recently they have added an online Gaian Congress in winter.

I attended the 2022 Fairy and Human Relations Congress outside Salem, Oregon and interviewed a number of people about their ways of communication with nature spirits of all kinds. I resonate with their description of their gatherings:

> There are many worldwide efforts to promote peace between peoples, races, and religions; and for the protection and preservation of animal and plant species. Here at the Fairy Congress we are among the vanguard of humans working to bring peace and understanding

between humans and the fairy realms. At this time of multiple crises affecting humanity we feel it is very important to seek alliances with as many light forces as possible in other realms. Mother Earth and the fairy realms are big players in what is transpiring on the planet. Though many deny their existence, the numbers of people who are tuning into the spiritual realms and the nature spirit/fairy/devic realms is increasing by leaps and bounds.

The in-person Fairy and Human Relations Congress and online Gaian Congress are great places to hear from and meet people from all over who are experienced at communicating with nature spirits. And there are always people there who are beginners, curious and wanting to learn more.

READER'S QUESTION FOR REFLECTION

Bring to mind your ancestors from many generations ago. What might they have been like as "tribal people who were spiritually alive and intimately in love with the natural world"? What might they have known about nature and her spirits, to pass down to you via genetic memory?

CHAPTER 4

WHY DO WE NEED TO HEAR FROM NATURE SPIRITS NOW?

Learning to listen to nature spirits, we can begin to understand what we can do to best support nature's needs.

What does nature need?

Nature has been growing and connecting and supporting all her beings throughout time. When things change, whether they are large or small impacts, she figures out how to create more life. Somehow. Regenerate—create more life—nature knows how.

Here in western culture in the United States, nature needs us to assist her in creating more life by stopping our habitual behaviors that destroy life. We humans have focused on human needs for so long and somehow have forgotten that everything we have comes from nature. Everything we make comes from the earth, the living planet. What can we do to encourage life?

The modern times of technology and more and more convenience have helped us to forget. We humans have designed systems to help us forget that nature's genius brings us life. Nature's genius is the source of human genius. And we humans are also part of nature. We are made of the earth, the elements and molecules of earth. Nature is within us as well as outside us.

What if we consciously modeled our lives on nature's genius?

In current times, this is often called biomimicry: seeing how nature approaches solving a problem similar to one we are having and mimicking her approach. Wonderful inventions and discoveries have come forth in this field.

What if we all consciously allied ourselves with nature?

We each are the result of an unbroken chain of life from the beginning. We all come from long-ago ancestors who were allied with nature. And we can learn how to ally ourselves with nature in our current time.

The current descendants of Indigenous people are a living wellspring of ancient connection to nature, and my western culture has a lot to learn from them. Where I live in California, Indigenous people explain that they have been given the responsibility by Creator to care for the land and all beings. They recognize all beings as sacred relatives. Indigenous people lived with nature in this way for millennia before settlers arrived. Valentin Lopez, chair of the Amah Mutsun Tribal Band in my region, says:

> We look at all of the things that we see—trees, grasses, clouds, flowers—they're all relatives. They have the same mother and the same father as we do. Father Sky is Creator, Mother Earth is our mother. That's what nourishes us. That's what gives us life. That's what brought us here. They are our relatives, and we have a relationship with them that is very loving, that is very caring and gentle.

We all can ally ourselves with nature at this time. My favorite ways to do this are to spend time in nature with respect, openness, love and joy; listen to nature-loving people who inspire me; and learn to communicate with nature spirits.

This poem is one I wrote about why humans need to hear from nature spirits:

SPIRITS OF NATURE ARE PREPARING ME

The spirits of nature want to talk with us
To work together with humans
To protect and restore the places where we all live

My human brain thinks I could learn unique solutions
By listening to nature spirits
By opening myself to the mystery further and further
But so far, most of the messages received
Have been about my own development, my own behavior

About removing my blocks
My judgment of self and others
So I can flow, like the water

About allowing myself to be fully me
To just be me
Rather than someone else's idea of who I should be

About knowing that now is the time
For me to come forward
For me to break free
For me to take action with a heart full of love

The spirits of nature are preparing me
To live in full interconnectedness with all life

I begin by experiencing the awe and wonder of nature
Open myself to new ideas and ancient practices
And merge with the mysteries that surround us

INSIGHTS FROM INTERVIEWEES

I interviewed a number of experienced nature spirit communicators for this book and include excerpts from these conversations and their teachings at the end of many chapters. Find out more about each interviewee in Meet The Interviewees.

Barbara Thomas

The gnomes say, "We cannot go on any longer alone. We are in deep need of the nourishment we receive from the interaction with the human kingdom. And you are in deep need of us, to sustain your life by interacting with nature. The simplest request we have is to recognize the life force within the soil, air, water, and the green and growing things…to recognize them as living beings and speak to us, relate with us."

Nature needs us to use our free will to ask for spiritual help. It is time to remember we are part of nature. We could not live on earth without nature. Nature needs our help to keep the air, water, and earth pure.

Michael "Skeeter" Pilarski

These quotes are from the introduction to his 2022 class, Ecosystem Restoration with the Fairies:

How can humans and faery beings work together to heal the planet? On the one hand, it's sort of the human hubris to say, "Oh we'll heal the planet." It's actually more like, "We can assist the planet to heal itself."

There's a lot of work to do to assist nature. But hardly anybody is working with the faery realms to figure it out. The fae are always working for healing. So, if we can collaborate with them, cooperate with them, we can get there alot quicker.

I remember a story of Ouapiti Robintree, our dear sister who taught flower essences many times at our event and passed on several years ago. Tacoma, Washington had a big copper smelter for many years, and the fumes blew across Puget Sound and landed an awful lot of pollution on Vashon Island and Maury Island. She was hired to go to Vashon Island to help them figure out how to remediate this heavy metal pollution. When she got there, she connected to the devic realm and asked them, "What could humans do to improve the situation?" And they gave her practical information to pass on to the restorationists, and the scientists, and the workers doing the work. She *asked nature* what they could do.

What we need: every restoration project in the world should have a psychic or a nature communicator who can tune in to the faery and devic realms and get information on what to do to help the situation. They have a deep, deep understanding, and they can give us really good information. We are the contingent of people who are working on it, and occasionally you bump into humans who are extraordinarily good at it. When you can tune in to nature, ask for advice, and carry the advice through to human language—that's something that I encourage all of us to do, and humanity as a whole.

READER'S QUESTION FOR REFLECTION

Why do many people find themselves drawn to nature at this moment in time?

CHAPTER 5

ENERGY HEALING AND RECEIVING MESSAGES IN THE FOREST

ENERGY HEALING

In 2009, I was deep into spiritual learning, which I have continued to study for many years. In my work at a spiritual center, my classes and practices led me to feel deep inside that I am a healer. My singing was healing, touching people directly at the heart level. My ability to listen deeply to people as a friend or coach was healing, hearing their truths and supporting them in their journey. I can access my ability as a patient advocate, accompany people to medical professionals in their time of vulnerability, and use my knowledge of science and western medicine to be assertive with doctors and get the information my client needs to make clear decisions, or know how to take care of themselves with recommended treatments. The word healer kept coming into my writing about who I am. I was accepted into medical school in 1976 and decided that was truly not my path, but rather my parents' dream. It wasn't going to be the western medicine path for me. And still I felt the call to heal.

I saw a film called *The Living Matrix* about current research and practices of alternative healing. The first section was an interview with

ENERGY HEALING AND RECEIVING MESSAGES IN THE FOREST

Eric Pearl, who founded Reconnective Healing, an energy healing modality. His patients had amazing healings as he passed his hands around their bodies, facilitating the presence of universal healing energies for them. There was the story of a young boy with cerebral palsy and his mother. The boy was able to move and use his body so much better after Eric's sessions with him. The boy's mother was surprised that Reconnective Healing was something that she herself could learn. She was trained to do it, as were many people all over the world.

I was drawn to this type of energy healing and began training sessions in Reconnective Healing in California. I appreciate that the healing that occurs could be at any level: physical, mental, emotional or spiritual. The level of healing needed at the time is a collaboration between the universal healing energies and the person desiring healing. It's not up to me. With a heart full of love, I am a facilitatior for the presence of universal healing energies for the person who is with me.

When I am doing Reconnective Healing, my hands tingle. I can feel the energy being directed through my hands; the tingling is my physical manifestation.

Although energy healing is not part of traditional western medicine practice, there are many scientific studies on this topic, documented in the film I had seen. This felt right for me, meshing my science training with a more expansive type of healing.

After being certified, I offered Reconnective Healing to friends, had several paying clients, and offered it for free to women in my community who had cancer. It can also be done at a distance, and I provided energy healing from my home for people in a hospital nearby and even people thousands of miles away. It can be used for animals as well; our cats, friends' dogs, and horses all received my energy love.

Once I started being in the forest often and connecting with the trees as sentient beings, I offered energy healing for some of them as well. It is a beautiful connection of universal healing energies, the beings, and me.

RECEIVING MESSAGES IN THE FOREST

For our Tune In to Subtle Realms of Redwoods forest experience, Dawni and I include a guided meditation under the redwood trees as our guests lie on a ground cloth. This is a way to connect deeply with each of our senses, focusing on one at a time.

Previously during our forest walk, we had spoken with our guests about the surprising fact that trees communicate with each other through their root systems. Forest ecologist Suzanne Simard at the University of British Columbia has been working on tree research for decades. I have been fascinated and excited about her work since 2017, when I heard her speak at the Bioneers Conference. Not only did she prove the ability of trees to communicate through their roots, she also proved the symbiotic collaborative relationship with the massive internet of fungi underground (mycelia) connecting root to root, sharing nutrients, water, and chemical messages with each other. In 2021, Simard published her story in her popular book, *Finding the Mother Tree: Discovering the Wisdom of the Forest*. We find that many of our forest guests are now aware of the trees' communication through their roots and the fungal networks.

So, when we focus on touch during the guided meditation, we ask guests to put their hands on the redwood duff on the forest floor below them. Then we suggest they put their fingers an inch deep below the top layer, into the soil where the roots of the redwoods live. And we invite them: "Feel free to ask for a message from the

ENERGY HEALING AND RECEIVING MESSAGES IN THE FOREST 41

Fingers in redwood duff while asking for a message
photo by Dawni Pappas

trees, here in this place where they can communicate with each other. The message may come in any form: words, or a vision, or a color, or a feeling, or any other way. Be open to receiving their message to you."

When I put my fingers under the duff and ask for a message, I feel my hands tingling. This is a familiar feeling for me, from energy healing experiences. And I often receive a message, often in a few words, sometimes as an image.

A few of the messages I have received:

- Flow
- The time is here. Allow
- Strength, flexibility, connection, community
- Deep roots
- Share love. We share love. The healing is mutual. The love is mutual
- Ceremony. Partake of my body in ceremony
- Criss-crossed lines, like a weaving that is not very tight (visual image)

The messages I've received are often directions for my own behavior or approach to life; that's how I interpret them. They are about opening deeper to the interconnected reality of life.

When we are with guests in the redwoods, when the meditation/sensory experience is complete, we ask if anyone wants to share what they experienced. "And did you receive a message from the trees? What was it?" A majority of people are willing to share the message. Many are surprised that they did receive a message; others are familiar with messages from nature. We often all share our messages, and sometimes share with each other the meaning it holds for us. Sometimes people receive transmissions from their ancestors, beginning a process of healing for them. Others seem not to get a message at all, or they receive a message later. Whatever happens is okay. We have created a container to hold the feelings and new awarenesses that often occur.

This learning journey I'm on has been a back-and-forth with me and the forest, an iterative process. It's like a spiral that began quite small and continues to fan out into a larger spiral. I go into the forest and have experiences. I go home and reflect on these experiences. I internalize what I've learned. These forest experiences often come forth in my weekly writing group, or my exploration with oracle cards, or my morning spiritual practices. Then I go back into the forest, feeling called by the trees. And the process begins again. I start from a different point on the expanding spiral. And the spiral continues to unfurl.

For me, these experiences with messages from the forest have been the spark that ignites my current process of deeper exploration into communicating with the spirits of nature.

Being with the trees was already quite familiar to me. It's just that I never consciously asked for messages before. Now I'm learning more about how to get myself into a state of love, respect, and openness so that I may receive more messages and take appropriate action related to this communication. I'm learning to understand what I can do to best support what nature needs.

I wrote this poem after we took a guest into the redwood forest to open and tune in. There in the redwoods, she had a powerful ancestral connection and healing experience. I'm grateful that she gave me permission to share this publicly, with the names changed:

GRANDDAUGHTER OF THE HEALER

The forest is misty this morning
No raindrops remaining from the light shower an hour ago
they turned into a magical place for sunbeams to enter
Tiny drops of water land on my nose, chin, and hair
pausing on this living surface, then coalescing into wetness

As Thea, Dawni, and I walk further into the forest
we can't help but feel our hearts open gazing upon such beauty

I sing us into the forest depths
the calling to re-wild our selves
Dawni guides us to touch the tree bark, the green fern
the thick oak leaf, the smooth red manzanita trunk
We stop in front of the redwood
stroke her leaves by the handful to drink in her aroma
We taste a few redwood leaf needles
the young light green ones at the tip
their astringent citrus flavor explodes against our teeth
Douglas fir calls out to be included
We slip fir needles into our mouths and chew
releasing flavor of grapefruit
Touching, smelling, and tasting deep into the beauty

We enter the redwood fairy ring
called "the sisters circle" by Thea's aunt
the women in her family quite familiar
with forests and their lifeways
Her grandmother Althea was a spiritual leader and healer
in the traditions of West Africa
who used herbs and hands-on healing
along with bones and fur

We lie on the forest floor connecting spine to earth
put our fingers under the surface leaves
where the dark damp soil has formed just an inch below
We feel the root energy there
and ask for messages from the trees

Thea feels Grandmother Althea's presence
at the base of the biggest tree, burned black
Grandmother Althea died suddenly when Thea was only four
Sadness permeated every room for years and years
So much grief throughout the family

they couldn't bear to even speak about Althea
Althea wants her granddaughter free in her own power
Dawni offers Thea our healing practices
Thea says yes, and we begin
surrounding her in love and healing energy
to release sadness, guilt
anything holding her back

Thea speaks her grandmother's name
Althea   Althea   Althea
Here in the forest Althea and Thea are brought together to heal
to connect through the veil
to share wisdom and love right here right now
Today is Althea's re-birthday
Thea is the candle
light streaming from her eyes, heart, solar plexus
full of Althea's love and permission
to be who she truly is
Gift of the Spirit

"GRANDMOTHER REDWOOD WANTS TO TALK WITH YOU"

Once as I was preparing to visit the redwoods near Oregon, I met someone who loved these trees deeply. We talked about our connection and communication with the redwood tree spirits. They closed their eyes and said, "Let me check in with my redwood friends there." And after a couple minutes, they told me, "Grandmother Redwood wants to talk with you." I felt honored and excited! This conversation with Grandmother Redwood would lead me to a key message from a redwood tree spirit about my communication with nature spirits.

A couple days later, my partner Ellen and I are heading toward

Oregon, driving to the location my guide had described. We park the car, look around, and I say, "Yeah, I think it's across the road over there where we begin." We cross and are looking down, where another paved road is coming up. It doesn't feel right. We're standing there trying to figure it out, and all of a sudden this young man in a jogging outfit comes running up the paved road. He comes toward us and says, "You seem to be looking for something. I'm wondering if I could help you?" Describing certain landmarks, I say, "Yeah, we're trying to find how to get to a path that takes us to bigger redwoods." He says, "Oh, that's right across the street, that's where I'm going. Follow me." We cross, and he runs up the path. He was right. This was the right path, and we never would have found it if we had gone that first direction. A faery guide, for sure.

We go up the path to find a sharp turn and all the other markers that had been described to me. I had learned that Grandmother Redwood had many different trunks coming up together. We definitely found her. And then Ellen went on up the path beyond because she knew I wanted time alone to connect and see what Grandmother Redwood might have to say.

I greet Grandmother Redwood with love, thanking her for her presence through the years. After giving her an offering of water, I ask permission to spend time with her and connect. Hearing *Yes*, I send her my love and touch her tenderly. I lay my cheek on her bark and sing verses of my song, "Sacred Forest," to her. I walk around her expansive girth, touching her and seeing her beauty. I ask if she has a message for me this day, but nothing is coming. I say to myself, *Well, I think I'm gonna put my hands in the soil where the roots are, where I ask for messages in the Santa Cruz redwoods.*

I go down to the path below her because I know her roots are

ENERGY HEALING AND RECEIVING MESSAGES IN THE FOREST 47

Coleen on the path near Grandmother Redwood
photo by Ellen Farmer

right there. I lie on my back on the ground, placing my hands on the duff and fingers down in the soil. Breathe deeply, in and out. And just like in the Santa Cruz redwoods, my hands start tingling… that familiar feeling. I send Grandmother Redwood my love and ask her, "What is your message for me this day?" Keeping myself present and open, I listen with all my senses for words, sounds, colors, feelings, images. I continue to send love and stay focused. The tingling in my hands increases. And then I receive the message: "Your hands tingling IS the communication." I receive a feeling of love and deep caring from Grandmother Redwood as she continues: "Stay open. Continue to connect. It will become easier." *Oh my gosh. Whoa!* And I just start crying.

This means that whether or not I get a message in return, when my hands are tingling as I intend to communicate with nature, I KNOW that yes, I am communicating. That was what I came here to learn. No one else could have told me this. Grandmother Redwood gave me this message that I truly needed. I am so deeply grateful.

READER'S QUESTION FOR REFLECTION

When you go to a favorite place in nature, have you experienced any sense of communication with nature or her spirits, perhaps a feeling, or word, or vision? Close your eyes and recall that experience now. Make a few notes, if you feel called to do so.

CHAPTER 6
SACRED FOREST SONG

In spring of 2022, I took an online songwriting course with H.E.R. I had written one song in the past but felt the call to bring forth my authentic expression in song at that point in my life. It's not surprising that the song I wrote was about my experiences in the forest.

I recorded the song with me singing and my friend Rhan Wilson playing the instruments. The recording feels so true to who I am at

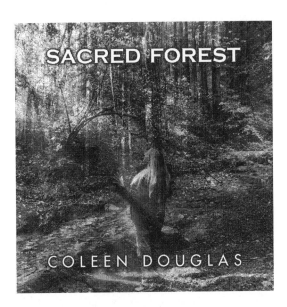

Sacred Forest song cover art
photo by Dawni Pappas, cover art by Rhan Wilson

this time in my life. I invite you to search online for "Sacred Forest Coleen Douglas" for the YouTube video, which includes photos in the redwood forest and a link to download the MP3.

SACRED FOREST

The forest is my sacred place
She holds me when I'm blue
I see her beauty, I feel her grace
And I come away renewed, come away renewed

Trees standing 'round me strong and tall
Devoted to their kin
Their roots connected, each one to all
They say I'm always welcome in, always welcome in

Walk slowly touch their trunk and leaves
I feel tree energy
They're my ancestors, here for centuries
Telling me to just be me, to just be me

I hear birds, and water sings along
Calling me to life
Nature spirits, they lead me on
Into wonder and delight, wonder and delight

The forest is my sacred place
I'm filled with gratitude
For your wisdom, your sweet embrace
And I come away renewed, come away renewed

Soon after the recording and video were completed, I received an email about the first-ever online International Conference on

Forest Therapy: Healing with Nature, July 7-9, 2022, organized by the University of British Columbia Department of Forestry, and registered to attend. The speakers came from Japan, China, South Korea, Colombia, Europe, Canada and the United States. I thought "Sacred Forest" could be a wonderful addition to this conference, perhaps providing music during breaks. I wrote to the conference organizer, who said, "I feel your song is so uplifting and healing. It feels like a wonderful hug full of nature's love." It became the theme song for the conference and was played twice at each break, with lyrics and my email address posted next to a forest picture. After the conference, the organizer thanked me and said "…it certainly did set the mood we needed to come together as a collective." One attendee asked permission to translate the lyrics into German and posted the results. I felt that the synchronicity of the timing for this was guided by nature. It was a joy to imagine my song circling the globe.

I know that one reason I am here on the planet is to be a singer. Singing gives me such joy! And singing is an essential way that I use my voice to touch human hearts.

Singing is also an important way that I connect with nature spirits, in the forest or during my spiritual practices at home. They seem to like it when I open myself to them in this authentic way. You might want to try opening yourself up to nature spirits with singing, instrumental music, drawing, dance, yoga, painting, or whatever form of expression feels authentic to you. Be willing to allow the fullness of you to shine through.

READER'S QUESTION FOR REFLECTION

What form of expression feels authentic to you, to assist you in opening yourself to nature spirits?

PART TWO

THE 8 TIMELESS PRINCIPLES

1. RELEASE JUDGMENT
2. RESPECT
3. LOVE
4. ASK
5. LISTEN
6. TRUST YOUR INTUITION
7. IMAGINE
8. CULTIVATE RELATIONSHIP

As a beginner in communication with nature spirits, I thought there was a secret to receiving messages from nature, and once I learned that secret, I would be able to communicate easily and effectively. So I decided to interview people at the Fairy and Human Relations Congress. If anyone could enlighten me on the subject, this yearly gathering would be the place to find them. I also consulted with four experienced nature communicators outside the conference: Barbara Thomas, Brenda Salgado, Kami McBride, and Pam Montgomery.

I discovered there is not one secret that everybody knows. Each person has their own way of doing it! As you read their stories, you will get a sense of the wide variety in how people communicate with nature spirits.

This section describes the 8 Timeless Principles for nature spirit communication, gathered from my personal experience and from interviews with seasoned nature communicators. Try these out to find ways to connect with nature spirits that feel right to you. Start there!

CHAPTER 7
PRINCIPLE 1: RELEASE JUDGMENT

This principle came to me quite awhile after the other principles, and takes its place as Principle 1 because it impacts all the others. Barbara Thomas told me: "Judgments shut doors and close hearts."

My judgments preordain what I am open to. I need to release my judgments to open to new experiences. I have held myself back from mystical experiences many times, thinking they were just too out-there, too woo-woo (though I've lived in California my whole life). Most of the time my judgments are knee-jerk reactions to an idea. When I look more closely, they are old and dated thoughts or fears coming to the surface, impacting my ability to try new things.

Learning to communicate with nature spirits has required me to release my judgment, over and over again, to get out of my own way.

Initially, I was far out of my comfort zone with nature spirits, but I loved being in nature so much and felt so good when I allowed myself to be there fully, with all my senses engaged. As my deep connection to trees had continued since childhood, if any being had a message for me, I felt it would be the trees. And then…I began to ask for messages from them, and actually receive messages from them.

Now, I continue to find myself releasing my judgments as I encounter the remarkable diversity of nature spirits and the people who connect with them. I don't want to be held back from this exploration. It's fun and magical. And it is deeply meaningful.

I do believe that humans and nature spirits can co-create a new earth together. And now is the time. So how do I release my judgments to remain open to new experiences?

RECOGNIZE WHEN I HAVE A KNEE-JERK REACTION

When I dismiss an idea or a person out of hand, before even giving it any thought, I help myself stop and reconsider: *Don't react how I usually react; see it in a different light.* Recognizing my own reactions is a practice. It requires repetition; meaning, I am not perfect, and this will keep coming up. Instead of allowing these reactions to control what I do, I recognize the situation and make a conscious decision about how to proceed.

QUESTION A HABIT OF RESISTANCE

When the possibility of a new experience comes and I find myself in a state of resistance, I question myself: *Why am I resisting this idea? Why wouldn't I go ahead and try?* There could be a good reason for my resistance. I need to use discernment to choose which way to go, not just give in to a habit of resisting. Dawni Pappas says, "When I let go of resistance, something magical happens."

MOVE TOWARDS LOVE

Granddaughter Luisa came to our house and wanted us to watch the animated film, *Strange World*. Immediately I thought, *Oh no, another animated kids' movie, I don't feel like watching that.* Luisa said, "I think you two would really like it!" Rethinking, she said, "No, I know that you two are really gonna love it!" Turns out we did love this creative movie about humans' relationship to nature.

And I could have just let this opportunity pass me by. Luisa knows and loves us, as we love her. Let love lead the way and open doors.

TALKING TO MYSELF

Watch your thoughts Coleen
Which ones are moving you toward love—and which ones are not?

If not moving me toward love, where are they moving me?
Toward habits
Toward fears
Toward media manipulations
Toward separation

I want to move toward openness
Toward confidence
Toward wisdom
Toward interconnectedness

Move where you want to be, Coleen
It's your choice
Your awareness

Keep taking action based on love
Keep allowing love to stream out from your heart
Your love energy has impact
Your love energy spirals out wider and wider

Share love energy with others along the way
With people, trees, animals
With water, ancestors, all nature spirits
Love is the power that unifies
Your love energy has impact
Let it flow

I know I want to open to the new experiences that nature spirit communication offers me, so I refuse to let old habits or fears stop me. Onward!

READER'S QUESTION FOR REFLECTION

Have you held yourself back from new experiences because of judgments about other people or worry about how others might think of you? As you consider an example in your past, do you wish you had taken the risk to do it? Why or why not?

CHAPTER 8

PRINCIPLE 2: RESPECT

When you go into nature with the intention to connect and communicate, it's important to begin with respect. Respect for nature herself. Respect for all the beings, seen and unseen. Respect for the Indigenous ancestors of the land you are on.

HONORING

One way to show respect is to speak to nature, preferably out loud, but silently works as well. Tell nature why you are there, your intention, and give thanks. An example:

> *Hello nature, good morning! I am here today to open to you and to connect deeply. I am so grateful for your presence in my life, and so thankful for all you give to the inhabitants of the earth. I honor your inspiring beauty and intelligence.*

Herbalist and plant communicator Kami McBride suggests that you may also want to honor the pain that nature has experienced. For example:

And I honor the pain and destruction you have lived through. Please forgive us humans and guide us in making the changes that are needed.

ASK PERMISSION

It's a good practice to ask permission of nature to enter a new location. "May I come in?" Then listen with all your senses for a *yes* or a *no*. Sometimes a *yes* can be subtle; if the answer is *no*, you will feel it.

If you want to connect with a particular tree or plant, to touch it, ask permission. "May I touch you?" Listen for your answer with all your senses. Say thank you.

To connect more deeply with a particular tree or plant, you may feel an inclination to taste it, to taste a leaf or petal. **Please research this first to find out if it's safe to eat**, as certain plants are poisonous. You definitely want to ask permission as well. "May I take one of your leaves to taste, to bring into my body, to connect with you in another way?" Listen for your answer with all your senses. Say thank you.

When you come back to a particular location over and over again, the beings get to know you and you get to know them. After a while, you will know intuitively whether or not you have permission. And you can ask them for more guidance: "How can I be respectful?" or "Please help me to know you." Again, listen with all your senses and say thank you.

Michael "Skeeter" Pilarski shared this in a class at the Fairy and Human Relations Congress:

> This is an Indigenous thought I learned…when a person arrives in a new land, they greet the land, and greet the beings of the land. They say "Hello! I'm asking for permission to step foot here, and to work with you." So

they acknowledge, they respect, and they ask. That's something we can all do—check in with the beings when you get to a new place.

You don't need to say oh, I'm too lowly, I'm just a human…no, you can talk to them. You can talk to anybody in the spirit realm. You may not hear an answer. It's easier to speak to nature than it is to hear the answers. Because we all have the capacity to speak. People develop the capacity to hear in various ways.

LAND ACKNOWLEDGMENT

All land was Indigenous land. Indigenous people have cared for the land and all its beings for millennia. And so much land has been taken away from Indigenous people by settlers, often under the most painful circumstances, including culture destruction, slavery, and death. This historical truth is finally being revealed. Many organizations and individuals are now honoring the Indigenous people currently living here as well as the Indigenous ancestors of the land.

Here is an example of the land acknowledgment we speak at the beginning of our Tune In to Subtle Realms of Redwoods experience:

> I give great gratitude and honor to the original people of this land—the Awaswas-speaking people, the Sayanta tribe, stewards of this land for thousands of years before the settlers came.
>
> I thank you for your knowing that everything here in nature is alive and is our kin. We are all completely interconnected.
>
> I give great gratitude for your descendants, the Amah Mutsun Tribal Band, who are relearning their earth practices and collaborating with many people and organizations to bring Indigenous wisdom into their land restoration practices.

You can learn about the Indigenous people on the land where you live by going to the Native Land Digital website. Native Land Digital, a Canadian nonprofit organization, has created and maintained this website since 2015. As they say, *It is not perfect,* but it gives a wealth of information, especially in North and South America, and Australia. You can find out more about Territory (Land) Acknowledgment on their website as well.

OFFERINGS

You may give offerings to the land itself, to the waters, or to specific beings you want to connect with. Any offerings should be natural substances, preferably native to the land you are on. You may give an offering of water. You may give an offering of a song or a poem. You may make a bundle of leaves or twigs that have fallen near you. You may make an arrangement of stones, shells or seaweed found nearby.

Your cultural background can certainly be an important source for traditional offerings such as tobacco. Although other common offerings are natural (like cornmeal, mica, flower petals from your home, fruit), they are not often native to the land you are visiting and may take longer than you think to biodegrade.

Offerings are personal. You will find an offering that feels right for you.

READER'S QUESTION FOR REFLECTION

What makes you feel respected? How might this inform your way of showing respect to the land, waters, and nature spirits?

CHAPTER 9
PRINCIPLE 3: LOVE

When you interact with nature spirits by consciously bringing forth your love energy and letting it flow out to them, they sense that you are likely someone who is open and sincere. Sharing your authentic love energy in this way is essential for connection. I wrote the poem below to express this vital truth:

LOVE IS THE OPEN DOORWAY

Love is the open doorway for connection
Love is the root consciousness
Love is the source which makes other things possible
Love is the power that unifies

Love is within our own hearts
We don't need to try to find it
Love is the open doorway, ready and available now
Our human heart center simply needs our focus,
over and over again

Practice letting the heart lead you in decisions and actions
Let your heart lead while your mind follows
When your heart leads and sets the standards
Your mind brings forth its powers in service to the heart
In service to love

Love is the open doorway for connection
Among people
Among people and nature
Among people and nature spirits

Open your heart wide
Allow your love energy to flow
This is your calling card

Love flows both ways
As we humans go into nature with a conscious opening of our hearts
Love flows out from us into nature
And in return love flows from nature into us
We feel the mutuality
We realize we are truly part of nature
We connect within the flow of love

HOW TO SHARE LOVE ENERGY IN NATURE

- Allow yourself to be in your heartspace, to focus on your heart center.

- Go into nature, physically or in your mind's eye.

- Give yourself time to be quiet. Breathe in and out several times. Let your body relax.

- Activate your senses. Look around and take in nature's beauty. Close your eyes and listen to the sounds of nature. Feel the air on your skin. Taste the air as you breathe in. Open your eyes and let your view help you feel a deeper connection inside.

- Bring forth a felt sense of love for all beings around you. Feel the warmth of love in your body. Be in your aura of love, your heartspace.

- Open your heart and send your love out. Imagine love flowing out from your heart, bathing all beings in love.

- Now you have let all beings know your intention is to connect in love.

- They feel it. And they will send their love back to you. Be open and aware to feel their love as you continue your connection experience. Love is the open doorway. Love is the power that unifies.

DANCE OF BEAUTY AND LOVE

Take yourself into your favorite spot in nature
Be still and take in the beauty
This is your kind of beauty
Touching you deep in your soul
Take a moment to feel that touch inside you
Where do you feel it?
What is it like?

My heart and throat are tingling
With the touch of nature's beauty
Nature reminds me that I am nature's beauty
"Let your beauty shine"

When I go out in nature and gaze upon her beauty
I feel my heart tingle, my breath deepen
Nature receives my beauty and tingles in her own ways
Knowing that I am here and we are collaborating
In love

It's so easy here in my favorite place in nature
To feel the impact of beauty and love
To pay attention as you take time
To watch, listen, feel, touch

Breathe again
Send out your deep love
Take in nature's deep love
Imagine them dancing together in the air you breathe

I see them approaching each other—
My love and nature love
Sparkling magic surrounds
They swirl around each other
Flying upward in spiral dance
Gathering more enchantment
As they rise together in joy
The curved shape of love
Planting seeds
In the opening portal

READER'S QUESTION FOR REFLECTION

What might it be like in the world if we all let our hearts lead, and our minds followed love's lead?

CHAPTER 10

PRINCIPLE 4: ASK

Ask nature to connect with you. Asking is your invitation for nature spirits to connect with you. The question I have asked in the forest most frequently: "What message do you have for me today?"

This seems like the most open question I can ask, which could be answered in any way by tree spirits or other nature spirits. I want to be available to what nature wants to show me and teach me.

Then I learned that I could ask for specific guidance from nature about many of my life's questions. Examples:

- What guidance do you have as I begin this new focus in my life?
- I'm feeling so much sadness. Please show me a path forward.
- What can humans do to help with the restoration of this particular land?
- Mother Earth, how can I be in service to you?

INSIGHTS FROM INTERVIEWEES

Stories from my interviewees broadened the possibilities for this principle of asking. Enjoy their descriptions of the variety of ways people ask nature to connect with them:

ASK FOR RELATIONSHIP

I took a course called Plant Wisdom from herbalist and plant communicator *Kami McBride*, who expressed that we are asking for *relationship* with the plant beings:

> How do you ask? Give it conscious attention. We're learning how to tune in to other living beings. You can ask: What is the best way for me to know you? Please help me to know you. Please teach me to learn from you.

Zing Nafzinger describes her first attempt to connect with gnomes when she wanted to learn to garden in harmony with earth's invisible forces:

> My first step was to consciously invite the gnomes to communicate with me. As I do when I meditate with my angels, I quieted myself, focusing on my breathing till my body settled. I felt my body relax. Communicating my intention clearly to the gnomes, projecting my thoughts, I was patient as I waited quietly for what I might perceive. There was nothing. I was alone with my intention. I let it be and trusted that, when the time was right, I would hear, feel, or see some kind of response. I repeated this invitation exercise each day. A few days later, a response came.

When I asked *Barbara Thomas* for advice about how I may better connect with gnomes, she said:

> First, ask: "Guide me into how I can relate with you. I want to relate. Guide me how to make that real." Then, sit in nature and open your heart. Just send love. And then follow any imagination that shows up. Any ideas that come, follow up.

ASK FOR ASSISTANCE

When you're working on something, you can ask for assistance from nature spirits. This goes for any kind of work, gardening, writing, construction, electronics, art…whatever you're involved in. If you've lost something, you can ask for help to find it.

Ron Hays

I was at a retreat with David Spangler, and he told me the story of a good friend, a deeply intuitive person. At the time, his friend was a remodeling contractor. One day, underneath the sink, he was trying to get a pipe loose and it wouldn't budge. At some point he said, "I just need the plumbing deva to show up and get this damn thing loose." So the deva of plumbing showed up, did whatever it did, and the pipe just came apart. Since I was also a remodeling contractor, I thought—oh, now that's interesting.

At the retreat center that evening I was alone in the hot tub. Suddenly the angel of construction showed up and was just there. And I just said, "I never knew you were a part of this." It was basically letting me know: "Make me a part of your work." I began to cry. It was a very powerful experience.

And that story triggered this sense that there's more to the built environment than I ever really allowed myself to know. So I began working with that. I would go into a house and try to sense the vibration, the spirit of the house. And it varies significantly from house to house. For instance, in a house that hadn't been cared for, it was like the spirit of the house was asleep, or dormant, or barely there. Inside the house of a friend of mine, who is an architect, the house was just vibrating, because he has taken such good care of it.

Barbara Thomas

One of the first real interactions I had with elementals was being guided to a lost object. One day, I lost a book. And I thought, well, maybe the gnomes can help me find it. So I asked for help to find this book. And the thought came to me: "Stop yammering. Get up and start putting things away. And we can guide you to your book. It is easier to move a rolling stone than a stationary object."

What made me know it was real was that they said "Stop yammering…" I'd never heard the word *yammer*. There was no question. Because of that word, I knew I wasn't making it up. I got up and started putting things away. I was guided, and I found it after the third thing I put away.

Later in my life, my son lost his car keys. He asked if I'd seen them, and I said, "No. You can ask the elementals." And he said, "Oh Mom, you just tell me that so I'll clean up my room." But the first thing he picked up had the keys under it. Then he came back and showed them to me.

READER'S QUESTION FOR REFLECTION

What would you like to ask nature spirits about?

CHAPTER 11
PRINCIPLE 5: LISTEN

Listen deeply, by opening all your senses: sight, hearing, touch, smell, taste, love, feelings, natural knowing, emotions, third-eye visions, dreams. Receiving messages from nature spirits can happen in any way, and can be different for each person. You may receive your message right away, later in the day, or in the future.

The more I spend quiet time in nature and open myself to nature spirits, the more I receive their wisdom. I often take a pen and notebook with me into nature, to capture the experience in words or drawings. Here's what I wrote about my experience alone in a secluded forest location in early spring:

Lying down below the giant Douglas fir tree, I close my eyes. My hands go out beyond the ground cloth and I rest my palms and fingers on the ground. It's cold, I wear gloves. I feel the familiar tingling in my hands, arms and chest. I can feel the energy exchange with the ground with my hands. My hands alternate from relaxed to pressing down hard, as if I could travel through the soil with my fingers. I feel the connections among millions of fibers—energy, mycelia, roots, beings. We are all connected together. And together we understand how to live together, and to co-evolve and co-adapt together. They tell me:

Be strong and big and tall
Send your roots into the ground
For the magic
For the communications among all
For the collective action

And then in fall, on a full moon morning, I returned to that secluded forest location to ask about what's coming in my life, as I learn more and engage in service to Gaia. Lying on my back on the ground, this time the message was all visual images. I wrote about them briefly then drew pictures of what I experienced.

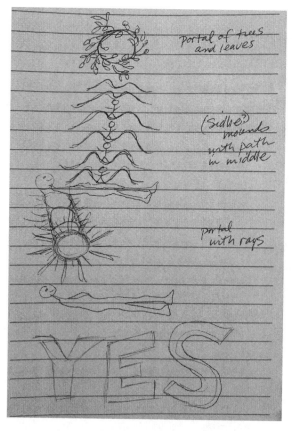

Journal drawings of visual image messages

Sun rising from my heart, opening
Bubbles rising up from my body
Portal circle of branches and leaves
Moving hill mounds on both sides of me, with a path in the middle.
The hill mounds would rise up around me on the path, then go down into the earth, as new hill mounds arose around me.
Portal in the air above me—a circular opening with rays of energy coming out around the circle, with a tubelike structure beyond the opening.
A big *YES* written on the sky

This poem is about my experience listening to the eldest tree on the land where I live:

LISTENING TO MAMA INCENSE CEDAR

I'm sitting on the soil created by you dropping your leaves
Over and over and over again every year
Sacred Mother Tree here on the land we inhabit
The eldest tree
The biggest tree
I honor you
I thank you for your presence
Here in this place for at least 80 years
Your beautiful trunk widening year after year
Now only six inches from the roof
We are happy to give you space
By removing the rain gutter you bent
When storms moved you back and forth
You are unconstrained
Free to be fully you
Mother Tree, what message do you have for me this morning?
I ask about developing my intuition
And my abilities to communicate with you and all nature spirits

Mama Incense Cedar on the land we inhabit

My back against her trunk
My hands on the ground
The tingling begins
Eyes closed
I feel small beings moving on my hands…ants?
Yes, a few small ants tickling me
Joining in the tingling feeling
I brush them off and soon feel them again
But this time they are nowhere to be seen

The message:
Be open to even the seemingly unpleasant beings here on the land

Will they harm you in any way, really?
Or do they have another purpose?
Are they curious who this huge being is
Who just plopped herself down on top of their pathway?
Do they feel the request for communication?

The ants tell me:
Nature is the largest and smallest
Elders and newborns
All are ancient ones
Born from their ancestors again and again
To be here on this day
Just as you are.

Crow caws above
Dog barks across the street
Ant silently walks her path
Cedar stands tall and regal
My back rests against your trunk
Thick bark with deep grooves
Spiderwebs catch fallen leaves
And small flying insect meals
Ants crawl into crevices in the bark near the ground
Meet each other on their paths
Taking care of community needs

We are all living and dying here together
There can be no other way

INSIGHTS FROM INTERVIEWEES

Enjoy the variety of stories from experienced nature communicators about how to listen and receive messages from nature spirits.

Camilla Blossom

For people who are starting out, communication with nature spirits can be whatever way it is. It could just be gardening with them. But really taking time to follow those inner promptings, inner sparkles that feel like *Hmmmm…what's going on over there?* And to see if you can engage.

Also tune in to how you feel. I do get some visions. Sometimes I just go places, like *I don't know why I'm walking over here…*but you do. It's just letting yourself be invited in.

I made up this term "seamless communication" because I realized as I was communicating a lot and learning more about nature spirits, that communication was happening all the time, but I just wasn't aware of it. I'm very empathic and sensitive, so just being in a forest next to a tree, I felt like I would merge in some way. My body would feel the energy of what the tree felt like…*oh I feel calm and grounded*. It helped when I started to teach, because some people want to communicate or see a fairy, or have a conversation where you use words. I recognize the language of nature is *feeling*…being in that state of feeling them and merging with them. Using seamless communication helped me realize how connected I was, already. And it's really helped some of my students to realize that they are connected, they just didn't know. You can be aware that Oh, *that's what's going on.*

I took Kami's online course called "Plant Wisdom" to deepen my connection and communication in the forest. Here are relevant notes from the course:

Kami McBride

Find your awe. Go where you're drawn on the earth. Think of a question you have for the earth. Walkabout for 30 minutes and be open to receiving your answer. Have faith. The earth is always communicating and something will come.

Receive. Let nature move you. Notice: color, or sun on leaves, or emotion, or smell. At that moment, you don't have to know what it means. The moment of the contact, itself, is the medicine. It is communion with the earth.

Focus on "open to receive" instead of "getting an answer."

Take time to reflect and describe: What came to mind? What's the mood? How did you feel? What's the energy? Pay attention to the exact details. What happened? Describe the process: Sudden flash? Unfolding? A concept? Pieces? Hear a voice? Was it like a movie? What sense or direction did it come in? Sometimes it's in the subtle details.

Dawni Pappas

I would describe my communication with faeries and nature spirits as "relational ancestral." I know, in every cell of my being, that I am a part of nature and they are my family. Since I was a little kid, my mom told me, "Mother Nature loves you, she will never harm you. She will always direct you." I've seen, tasted, touched, dreamed of, believed in, built homes with, and guided and guarded my relationships with faeries and nature spirits my whole life. Trees and grass talked to me. Thunder and lightning were my friends too. I thought everyone knew that there were nature spirits, and that trees and grass were living beings who talk to us.

I hear, I see, and I feel messages, and then I'm guided. How I receive the message depends on where I'm at, mentally, physically, and spiritually. If I'm focusing on a question, I will tune in. And it's like a prayer, you just allow yourself to be open and receive, and then trust the answer, or the feeling, or the nudge, because it doesn't always come in words. It comes in feather touches, or a gentle movement towards a direction. Or my favorite, a release of what is holding me back.

Part of getting my messages is actually putting them into practice, because I don't want to stop the flow. We are nature, naturing. I am a nature spirit. I believe you are too. So how we communicate with ourselves and take our own divine guidance is really important.

Summer Starr

Most of my communications are auditory, like having a conversation but in the head through telepathy, using words. But sometimes there are impressions of visuals that can come in as well. And when there is enough trust between the two, there can also be somatic impressions as well, where I can feel the being energetically.

Brenda Salgado

Everyone has different ways of receiving communication. For me, whether it's ancestors or spirits or animals, one of the things that happens often when a message is coming in, is my whole head will start to vibrate. This is just an indicator to pay attention—incoming, incoming—slow down and listen.

Often I experience it as a message—in my head I hear a voice, so it's words. Or it can be energies, emotions, or vibrations. I'll get feelings from ancestors, and other spirit beings too.

I have lots of friends who see spirit beings in different ways. For one friend, when her ancestor comes in, she can smell her perfume—some people get messages through scent. It's really beautiful to think about all these different ways, and that we all have the capacity to cultivate these sensitivities.

Lindsey Swope

My communications are very in the moment. I would say perhaps I'm *clairsentient*; I get feelings, and then I have to put them into words. The wind and clouds often talk to me the most directly. I'll put a question out, maybe it's a yes/no question, and then I'll get a breeze that will come and sort of affirm…Once I was going to a place on the Isle of Skye where the fairies of Scotland are very active, and I felt this wind stop me: "No, don't go any further." I could feel them checking me out…saying, "Who are you? What are you doing here?" I just stayed and was open, and then I felt, "Okay, you're good and you may go forward."

So, I would say I definitely feel. I don't often get words in my head. From a young age I've always felt called to be with nature, and that's how I ended up studying geology and geography. I've always been sensitive to energies, but it took awhile for it to become clear communications.

Pam Montgomery wrote the book *Plant Spirit Healing: A Guide to Working with Plant Consciousness*. She spoke during the Global Earth Repair Summit in 2022, at their online conference. I asked her this: "You say that plants' methods of communication with each other are light, sound, and vibratory resonance. But not words. So how do you receive your messages from plants?" Pam's reply:

Pam Montgomery

Light, sound and vibratory resonance. (laughter) Light and sound carry vibratory resonance. You can receive the resonance, the frequency of the wavelength. It's coming to you, and your body experiences it as a felt sensation. It could be heat, tingling, tightness…it could be anything.

Then you "put a handle on it," which means you describe it, just a couple words, enough so you can remember it. Just a few words, like, for example "soft as silk," or a posture, or it might be a symbol, or could even be a smell.

Once you receive the felt sensation, you know what the vibration of the plant is. Then you begin to build your relationship into a co-creative partnership where the plant spirit lives in your body or "your house."

Now it doesn't mean that there aren't going to be words. But the words are not the resonance. The words are the translation. The *resonance* is from the light, and the sound, that carries the vibratory resonance. The light and sound…that's the foundational way that communication takes place in the biological world.

Jazz

Black bear was a big totem for me. I had a black bear necklace, just one claw, that someone gave me. I wore it because it was a symbol of community and strength to me. When I moved from the eastern U.S. to the west, it got me through some really trying times. And then eventually I took the claw off the necklace and had this communal thing, where the black bear was like, "I'm ready to go back home. And when we go next time, leave me in the forest, let me go back to the earth. I'm still here with you, but you don't need this to have this connection. And you don't need me as much as you did. You don't need the lessons. You've already got the lessons. You learned it."

Messages come in different ways with different spirits. I've had potent feelings. And I've shared those feelings with other people. Like in a moment where they're very very present, those feelings are very evident, to the point where we both look at each other and say, "Did you feel that?" "Oh my god, yeah, I did." "Are you here with me?" "Yeah I'm here with you. That happened."

And then there's other times when it's just me in solitude. Especially now, recently, I get a lot more visions. Just very evident, very lucid flashes of things and imagery. But it used to be words, a little distant in a way, not as personal—but very matter of fact, very truthful.

Everything has something to teach you. I guess the basis of my practice is that I try to listen more, and in those moments I also try to create. I listen, and I receive, and then I give. I receive some understanding of nature around me and then I put it in a different form. You know, I didn't carve this stick here, only the bottom, because this had its own character. The same thing with that one. I gave it a little bit, I *saw* a little bit more into it.

Ron Hays

When I work with the tree spirits, the plant spirits, or the land spirits, I just usually stop, quiet myself as much as possible, and tune in to what's there. A lot of my awareness of other dimensions is through the felt sense, it's kinesthetic. So I don't usually have visual or auditory impressions. With this felt sense there's a translation of a sensation in my body into an understanding.

Aquamarine

Much of my communication is sensory in nature. I feel it. I hear it a lot. I'm auditory, so when I'm quiet I tend to hear things. It usually just occurs to me. Sometimes

I might see something or feel something, but because I have an existing idea about what I'm expecting to see or feel, I might not really follow up on the new idea, until I get a nudge to do so. The last online Gaian Congress was during the winter and I decided to go outside before the session started, to connect into my garden and yard. I know there are fairies out there, but I needed to go outside and check who was out there at that particular moment. In one part of the front yard, I kind of felt-saw a dragon. My reaction was, *I have to keep going.* But then later, *Oh, there's that dragon again,* which I had never noticed there before. I kept getting nudged to see the dragon. And later on, it kept coming up during the Gaian Congress session.

In her book *Faerie Wisdom to Live By,* Zing translated communications from a faerie named Celerine:

Zing Nafzinger

There are many ways of moving forward, and they all start with stopping and listening to those soft, light voices in your heart. Find a beautiful quiet spot and settle yourself comfortably there. Have paper and a pen where you can reach it easily.

Allow your body to relax. If there is any tenseness in any muscle of your body, give that muscle express permission to relax. You are lying in beautiful water, floating peacefully. Sense the soft inflow and outflow of your breath. You are in a place so beautiful that all your cares have disappeared. Now feel yourself being lifted up into the air so that you are effortlessly flying and can go anywhere you want.

Listen to the light faerie voices inside your heart. Surround your heart and these faerie voices with your love. Take your paper and write down exactly what you hear or feel, without judgment of yourself or the words. Be a good scribe and simply write down what you are experiencing. Don't think; just experience and write. You will know when your listening is complete. You will feel light and love, joy and playfulness in your heart. At this point, say thank you to the faerie voices and allow yourself to stay in that energy.

READER'S QUESTION FOR REFLECTION

Which descriptions of how people listen especially caught your attention? Which descriptions resonate for you? What advice would you give yourself as you practice listening to nature spirits?

CHAPTER 12

PRINCIPLE 6: TRUST YOUR INTUITION

Intuition is a direct knowing or an inner knowing, without thinking it through. Intuition is receiving information without using logical or rational processes. You can experience it in many ways: a gut feeling about a person or a situation, an unexpected insight, guidance received from a spiritual experience. It could be an inner voice. It could be something you just knew immediately.

Many of us in current times have not been taught much about intuition. People have been told they are silly for believing their intuition rather than their logical thinking. No wonder so many people second-guess themselves when they receive an intuition.

When I began to communicate with nature spirits in the forest, I would ask, "Do you have a message for me today?" and listen with all my senses. Then sometimes a word or a phrase would actually come into my mind! And I would tell myself, *Oh Coleen, you just made that up.* I would discount the messages I received because I couldn't believe it was genuinely happening, or I expected something else as a message, or I had no idea what that word or phrase meant for me. I thought I would get a clear message giving me specific direction from nature about what to do in my life.

For example, I received the word *Flow* as my message from the trees a number of times. Did that word come to mind because I could hear the creek flowing next to me in the forest? Was that really a message from nature spirits specifically for me? The more I began to trust that these messages from nature were real, and engage my intuition to interpret them, the more I could understand. Reflecting and writing about it provided more insights:

FLOW into your life, Coleen
Go where your passions live, and have yourself a life full of passion
Your passions are your gifts
They are what you are here to give
Flow is connection with ease and grace
Energetic flow
Feel the energy exchange between you and nature's beings
Allow yourself to do this
Be this

ALLOW is also a message I've received from trees
Allow yourself…
You, Coleen, need to let go, trust, and then
Allow yourself to be in your full sovereignty
To be what you are here to be
To give your gifts
To receive awe and wonder
To live life to the fullest, the deepest

INSIGHTS FROM INTERVIEWEES

How can we develop and trust our intuition? Like many aspects of communicating with nature spirits, it is a practice. We need to be patient with ourselves as we move ahead on this path. And then, one day, it will come with ease. These interviewees provide their experiences and examples that can be of assistance in our efforts.

Brenda Salgado

Intuition is really important. Especially because we're raised in the U.S., where we are taught to prioritize the mental and the thinking, and we've been taught to repress the intuitive and other ways of knowing outside of the five main senses. But intuition is innate to everyone. It's everyone's birthright to be in their intuition, their discernment, their listening, their connection, their grounding with mother earth. We've been cut off from ways of knowing and ways of being in relationship; now we're living in a time when we're reclaiming this.

I always tell people, when we're in different ceremonies or visualizations, that everyone is going to experience this differently. Don't second-guess that you're doing it wrong or right. Whatever is arising is what's right for you. Whatever you're intuiting is right for you. And that's part of developing your intuition—knowing we have this direct relationship with spirit, with our body, with the earth. And that it doesn't need to be mediated by some authority. It's a different way of knowing.

We need to be cultivating our intuition, trusting our intuition, and learning to listen. Also to be in discernment. Ask yourself, *Where is this message coming from? Is this mine? Is this someone coming to give a message? Is it of good intention?* You can set filters by saying, "I ask only for those messages that are of good intention, that want to support me on my path and purpose, and my health and wellbeing." If there's any other intention, you do not want to let that through.

It's important for people to be playful and open-minded to cultivate intuition, and to be patient as it grows. You know, if you're trying to build muscle on your arm, you don't expect to go once to the gym and have a big muscle. You go and you do repetitions, and at some point your muscle gets bigger and stronger. That's true around this kind of discernment and intuition too.

Kami McBride

Trust what you get—the inkling. What you received has something for you. Trust that it will provide a stepping-stone. What's awakened in you? Look for your unique way to express what you receive: music, writing, drawing, poetry, movement, dreams, song, craft, painting, talking with friends, anything. Don't compare yourself with others because that can shut you down. What is your unique integration process?

Dawni Pappas

I think intuition is total creative energy. How things reveal themselves to you, like a song, or a nature spirit, or a mantra—they do it in their own timing. The sculptor doesn't always say, "I'm gonna do this and that to this piece of clay." A lot of times the sculptor will come up to that piece of clay and say, *Show me. Reveal to me who or what this is going to be. Take me on that journey. I may not like it. I may not always agree, but I'll always come back, with harmony and balance.* That's your intuition. And the intuition doesn't come out just from you. The intuition is coming from the clay. The intuition is coming from your breath. It's coming. It's alive, very much so.

Sometimes your intuition will show you things or bring you places, and then the mind always wants to know what to do with it. But sometimes it's not about that. It's really about being. Intuition is part of being you and following that energy. Intuition is about following your heart.

READER'S QUESTION FOR REFLECTION

Think of a time you received some kind of knowing via your intuition. How did it come to you? Did you follow your intuition? Why or why not? Then what happened?

CHAPTER 13

PRINCIPLE 7: IMAGINE

Using your imagination can help you connect with the wondrous diversity of nature spirits. When we imagine, we open to possibilities that we wouldn't have experienced otherwise.

I am quite a visual person. I began nature spirit communication with trees, which I can see. But I've been learning that there are many other nature spirits, such as faeries and gnomes, that most people consider to be invisible. When I imagine what they look like, it helps me begin to connect. Over time, as I develop a relationship with them, imagining what they look like is less important.

In the Gateway to Gaia—Land Alchemy Training course with Camilla Blossom, we learned about earth elementals, including gnomes. I have a small gnome statue in my garden but I'd never had a personal relationship with a gnome. To connect with these earth elementals, first we relaxed into a connection journey. Then we did a "gnome dance" that Camilla demonstrated for us, slowly stomping one foot at a time in a circle, sending a blessing of love and gratitude with each step. (You can read a clear description of the gnome dance in Camilla's book.) I felt silly doing this gnome dance, but it turns out that gnomes prefer silliness over sternness in humans.

Then we stopped, with the intention of meeting our personal gnome. The instruction was "Use your imagination to *pretend* you see your gnome. Notice or sense what they appear to look like." I closed my eyes and imagined a gnome in front of me, about two feet tall, with a red hat and blue coat. ("No green coat," he said.) I bent down to be more at eye level, and started shining his shoes. When I got closer to his face, he said, "You have a big nose!" Looking back at him, I said, "You do too!" That made me laugh out loud.

The next day, I went out to the gnome statue in my garden. Plants were growing around it, obscuring it from view, so I weeded around the statue. Then I took a good close look at this six-inch statue and saw that the gnome was pointing to his nose! And again, I just laughed out loud. I'd had the small statue for a year but never noticed that before.

Gnome statue in my garden

A month later, I interviewed Barbara Thomas, who has decades of experience communicating with gnomes. After I told her that story, she said, "What I love about it is—how could you make that up?" Then she asked, "Did the statue pointing at his nose confirm your earlier conversation with the gnome?" I said, "Oh yes, definitely."

Dawni Pappas

Imagination is super important. It's an important tool because it's a felt sense that uses images, sounds, and the body. When we imagine, we also use our intuition—they flow together like an infinity sign or a torus. Imagination is possibility. Imagination releases us from limited beliefs and releases potential.

As we focus and call forth our imaginations, many times we release our needs about how things are *supposed* to be. When we do that, our imagination comes through seamlessly, as though volunteering—waiting for us to call upon an invisible friend, who is always alive in us somewhere, ready to play whenever we are. Imagination also brings forth creation. And in that creation and creativity, it disperses an unlocked power, joy, and responsibility. Giving permission to our imagination can be life changing in so many different aspects of our lives. And then, in the lives of others.

Camilla Blossom

A lot of people want to see fairies. "I want to see it! I want to see it, and then I'll believe it." And 99% of the time, that's not what happens. But you may feel them and may have telepathic communication. Maybe you just feel love when you're in your garden. And just imagine; imagination is a huge piece. If you're in your garden, you can ask, "Fairies, do you want to work with me?" And just use your imagination, and pretend that you can see them around, and ask them questions like, "Where does this flower want to be planted?" You gotta play with that… kids can show us how.

I wrote this poem in spring when hummingbirds frequent my backyard:

BECOMING HUMMINGBIRD

Three feathers guide me to the east, the rising of the sun
The energy of air elementals—wind, breath, air dragon, clouds
Angel wings move the air so potently, so gracefully
Feeling the slight movement of the invisible against my face
Air calls me to open
Open to the portal of love alchemy
Breathe deeply
Take in the love

Hummingbird flies next to me
Her body shiny green
Wings moving so fast
I can see the air move in her rhythm
She points the way to herbal magic
Rosemary looms large
Blue blossoms a reservoir of sweet nectar
She visits them one after another
Sipping their magic juice
She calls me to her side

Hummingbird tells me
To open to all possibilities with her
I grow smaller and smaller
To travel with her upon trails of air

I feel my body shaping into aerodynamic alchemical aspects
Smooth feathers emerge from each pore
Soft and blue and shiny
My nose and lips extend out gracefully in pointed beak

Thin tongue emerges quickly
Then rolls back inside into reshaped sinuses
Arms become winged
Beautiful feathers expand to carry me above

I am open
I am ready
I am flying
I am by her side
Sipping nectar from light blue blossoms
Rosemary, the herb of remembrance

I feel ancestral connection to the herbs
To the birds
To the magic of healing and flight

The Mutsun people of this land revered hummingbird
Who brought fire to their tribe during the time of floods
Fire through the sky for warmth, for cooking, for worship

Portal shows me a moment when I was a child
Who wanted to fly more than anything
I stood atop my toy red fire truck in the yard
Trees around me, leaves shaking with excitement
I opened my arms wide and moved them through the air
I saw myself rising off the earth
Propelled upward by heat of thermals
By movement of wind
Flying with the birds in complete joy

In the book *Engaging with the Sidhe: Conversations Continued*, a Sidhe woman, Mariel, has deep conversations with author David Spangler. Mariel describes what imagination is for the Sidhe, compared to humans:

For you, imagination is a subjective experience. For us, it is objective. It is a quality of fluidity and responsiveness within the life of the matter that forms the substance of our world. It is why our world is so directly responsive to our thoughts.

For you, imagination is an internal experience. For us, it is the substance of the world we live in…We are the ecologists of imagination; we know what it is in its natural habitat. We live with it in the wild, whereas you live with it in the laboratory of your minds.

We do not possess imagination individually as much as we participate in the "ocean of imagination" that is part of the life of the world. This does not mean that we cannot form images or imagine new forms as you do, but we do so in collaboration with the flow of imagination within the world. This is why it is very easy for us to manifest in form and substance what we think about, an ability that seems like magic to you. But your ability to fashion a complete world in your imagination seems like magic to us. Do you understand? It's as if you have your own private inner ocean that you can populate with life of your choosing. We do not have this. We work magic in our outer lives; you work it in your inner lives.

This is a power, my friend; it's a power to shape your outer world, certainly, but even more it is a power to shape yourselves for better or for worse. Therefore, if I have advice to give, it is first to be aware of how you are using this power. How are you imagining yourself? What inner world are you creating that then impresses itself upon the fluid matrix of the world around you? Second, use your imagination to connect. Use it to enter the inner worlds of others different from your own so that communion, connection, and understanding can build between you. Use your imagination to love.

I had to read this chapter on imagination in Spangler's book several times to try to understand imagination from the Sidhe experience. Imagination is a greater power of connection than I had realized. I wrote this poem after taking in Mariel's explanations of what imagination is to the Sidhe:

IMAGINATION

I've been trained to think my imagination is my own private world
I can make up and envision whatever I want in my head
Keeping it inside me, unaffected by the outside world

Nature calls me into her ecosystems to experience something else
She guides me to find spots where I can be in silence
Breathe in and out
Relax my body and mind
Feel my own boundaries and allow imagination to see them
Becoming permeable

The boundaries of my body begin to open
Like pores of my skin on a hot day
The energy field around me expands outward
Becoming wider and thinner at the same time
And wider and thinner again, like wisps of fog

Now nature's ecosystems can be
Seen, felt, and heard through nature's imagination
Flowing through my permeable boundaries
Into my energy field
Into my body
Nature's imagination is expansive
Her images, feelings, and sounds flow into me

Trees lighting up in delight
As sweet food travels down branches, through trunks and roots
Into the mycelial network connecting them underground

Grasses caressing sensuously
As breeze moves them
Ocean and sand dancing together in harmony
As waves break upon the shore

Flocks of birds in murmuration
Display the magic of individual movement
Shifting the whole community in joy
Nature tells me
This is the same power you have
To shift the human collective imagination
With the energy of your love

READER'S QUESTION FOR REFLECTION

What individuals or experiences could you imagine, to connect with nature spirits in a loving way?

CHAPTER 14

PRINCIPLE 8: CULTIVATE RELATIONSHIP

Have the intention to cultivate a relationship with nature spirits over time. You may not receive communications immediately, but you can always send your love and greetings out to nature. As you continue to open to possibility, nature spirits may come to you for connection.

SAY HELLO

For my birthday, I was inspired by what Barbara Thomas heard from her gnome friends: "The first request ever made of me, and the one I think I've heard of most often, is simply to say hello." I asked my friends on Facebook to celebrate with me by connecting with nature with this post:

> Today is my birthday! I feel so alive and energized by my work and play with nature.
> My birthday request for you is —
> Say hello to some plants today…a tree, a shrub, a flower…and tell them how beautiful they are.
> You could do this out loud or in silence…just send your love energy out to them. Plants are sentient beings and they feel it when you notice and greet them.
> Thank you, my friends! If you want to, let me know who you said hello to. Enjoy this day in a new way.

109 people liked or loved it, and 39 people wrote back that they said hello to various plants! Turns out that many of my friends talk to plants regularly, but for others this was a new idea. I loved hearing about how this simple action felt so good to them. And I know the plants and the gnomes were happy about it too.

DEVELOP ROUTINES

You can develop routines that will help nature spirits feel your presence regularly and get to know who you are. Barbara Thomas told me, "The more I experience and accept my experiences, then the more they happen. I'm open and they must know that. There must be a little sign: "She's open!"

Because I have been walking the same paths in a nearby redwood forest for many years, stopping to be with certain trees or the creek, I feel that the nature spirits in this forest know me. And I continue to receive messages from the forest beings, from the very simple to more complex communications.

I also have a morning practice of going outside in my backyard to be in the hot tub. There, I relax my body, experience the beauty of nature surrounding me, listen to and see the birds, consciously connect with nature spirits, sing, and let my imagination soar. I love the sacred waters surrounding me in this spiritual time.

Right next to the hot tub is the location of the land altar I created at the beginning of my Gateway to Gaia course last year. When the assignment came to "create a land altar in collaboration with the spirits of your land," I thought it would be located in the lush area we call the grotto. But, no, they wanted it to be next to the hot tub, where there was a rusty outdoor table and lots of weeds. After I cleaned up that spot and created my land altar there, it became obvious why. I am out there every morning in my sacred waters, and now I see and connect with the land altar and all it means to me every morning as well. As a result, I care for the land altar and keep that whole area beautiful and accessible.

In her Plant Wisdom course, Kami suggested specific awakening routines, letting all the beings know that you are paying attention:

Kami McBride

Hello Ritual: Begin and end the day with connection. Each morning, say hello to a plant you see every day, either from your window or when you go outside. Each night, go out and say goodnight to the stars.

Routine Offerings: For me, it was putting clean water in the birdbath and greeting the birds. You may want to make an altar in your garden for offerings. Or send greetings to your garden.

Barbara Thomas

I now live on the mountain in the heart of a redwood forest. The Amphitheater is a large open space surrounded by mature madrone and redwood trees. I've walked to the Amphitheater more than 6,000 times over the twenty years that I have lived here. My Council of Gnomes told me, "Open your heart, Barbara. That's the only way to see and know us. Your daily walking the same path has alerted the beings who live in the land of your presence. You have often been aware that they walk with you."

RIGHT RELATIONSHIP

Brenda Salgado is a first-generation Nicaraguan-American, born and raised in the San Francisco Bay Area. Trained by teachers in the Purépecha, Xochimilco, Toltec and other indigenous lineages, she draws on the healing powers of the natural world to guide her work. I asked her, "How would you describe your communications with nature spirits or ancestors?" Brenda's reply:

Brenda Salgado

It's an ongoing relationship with the beloved unseen. I recognize that I'm part of this larger whole, this larger family, that includes humans and the non-human too.

Grandma on my mom's side is an ancestor who watches over me and guides me in many different ways. She's connected me to different teachers. The very first teachers she led me to were elders from Michoacán and Xochimilco in Mexico. These were people with an unbroken lineage of medicine. In all of those cases, they didn't start teaching you medicine or ceremony right away.

They first taught you how to be in *right relationship*. A lot of attention was given to introducing yourself to trees and plants and mother earth, and to asking permission if you needed something. Not just to take the medicine, but to ask for the plant's advice. "I'm working with this person, and I'm wondering if you can give me some advice about how to work with them?" Then just sit there and listen to the plant. See the plant as an elder and a teacher, not just as something you're consuming.

I remember one of the teachers saying, "Even when you ask permission, you should wait a minute before you take the piece that you're going to take from the plant." It's really important to do that because you need to wait for a response. Otherwise, you're not really asking permission. He said, "You want to give a plant a second there, because what it can do is pull more of its life force into its core and push more of its medicine into the piece that you're taking. So the plant is less harmed in the process. The medicine will always be strong, whether or not you ask permission. But it will be stronger if you are in good relationship this way."

One teacher works a lot with white sage, and he's had relationship with some white sage plants for decades. He said, "One day I went to a white sage plant that I'd known a long time, and someone had overharvested to the point that they killed the plant. I wept at the side of that plant for hours. I need you to understand I wasn't crying because someone had killed a plant I harvested. I was crying because they murdered my elder, my teacher, my friend—who gave me so much advice over the years."

I understand why Grandma wanted me to start with these teachers. She wanted me to start with someone who taught right relationship before teaching medicine. I really appreciated that.

It felt really natural to me when I was young to be curious and listening and asking for guidance, asking for advice from the non-human world. And they're so patient and loving and generous when we do that. These have been old relationships. There are many trees around my house that I have deep relationship with. There are trees in other places that I have relationship with, that I've been in ceremony with. I've heard some native elders say, "We're the youngest relatives in the family. We have a lot to learn from these other beings."

Here is my story about an unexpected relationship with beetles who came to me again and again in different redwood forests.

BEETLES WANT TO BE WITH ME

In the familiar redwood forest near my home, I had an unusual experience with beetles that came to me each time I visited that forest. A few months later, traveling to northern California redwoods,

I was shocked to find the same kind of beetles coming to me, hundreds of miles away. They were persistent and seemed to want a relationship with me.

April, Santa Cruz, central California: Arriving in my secluded spot in the redwood forest, I send my love, and breathe in and out the trees. I feel the awe and freedom. I give great thanks for the forest's presence. I sit on the ground to begin, and ask, "What message do you have for me today?" Two black flying bugs come toward me and I shoo them away. One or two at a time, they keep returning. I don't want to be bitten, so I keep shooing. I notice four or five of them flying nearby, then alighting, a kind of black and coppery shiny beetle. I'm still upset and annoyed. *Should I leave here?*

What do they want? I think they might just be curious about me. I let one land on me, a three-quarter-inch shiny beetle with small black wing tips peeking out from the shiny carapace. Looking closer, I see the antennae. So small and feathery at the ends. The thin head, shiny carapace, glints of copper in the black. No biting. No problem. Just curious.

The message? Allow life to show itself, even if it's not what you're expecting. Be curious about the beetle—that's what the beetles are doing—they're curious about you! Be the love energy. Be the awe energy.

June, Branscomb, northern California: The beetles are attracted to me, the small ones with the shiny black-and-copper outer wings covering the lacy wings of flight. They tickle my skin as they walk and explore, one at a time, sometimes more. I've learned to relax and connect in. Why are we here together? Whether in Santa Cruz redwoods or Branscomb river edge, you find me. What is this relationship?

I select an oracle card from the Card Deck of the Sidhe for insight. The Sidhe say it's shapeshifting—their way of coming into attunement and blending with the life around them. The *Living Stone* card: "A standing stone is turning into a plant, perhaps a tree—or a tree is becoming a stone. In either case, *anwa* is being shared. Shapes are shifting, connections are being formed, attunement is taking place. And the result is new life emerging. Shapeshifting is not simply about taking another's form. It is about a connection that creates emergence and the sprouting of new insights, new beginnings, new life."

What is the *anwa*? The Sidhe say, "If we see a tree, we are aware of the spirit of the tree, and also of its form—its trunk and branches, how its roots go into the earth, the shape of its leaves, and so forth. The *anwa* stands between those two things: spirit and form. It is the shape of the tree's spirit's intent, and the activity and functions that flow from that intent. When we shapeshift, we use our *anwa* to enter into and hold within ourselves the *anwa* of the tree. We take it in ourselves, in our own energy field, the movement and flow of the tree's life."

Beetle's lacy flight wings beneath shiny outer wings

In this case, shapeshifting is connecting the *anwa* of beetle and me
For conversation, for heart-opening, for love

Part of the beetle's *anwa*
Are the shiny black-and-copper exterior wings
Hard and protective
With lacy flight wings underneath

Part of my *anwa*
Is a shiny protective covering
With the freedom of flight underneath

Here, shapeshifting, I take in my own energy field
The movement and flow of beetle's life
The message?
Open my protective covering, and fly!

READER'S QUESTION FOR REFLECTION

What is one being in nature that you have a relationship with now, who you would like to know in a deeper way? How might you cultivate that relationship further?

PART THREE

DEVELOP YOUR ABILITY TO COMMUNICATE

Now what do you do? Practice what you're learning and develop your ability to communicate with nature spirits. Go into nature and begin connecting with nature spirits by yourself or with a trusted friend. Open to nature spirits using the 8 Timeless Principles. Enjoy your time in nature!

These final chapters include a variety of possibilities for learning and practice:

GOING DEEPER: Go Into Nature and Connect; Discernment and Boundaries; Mentoring

ANCESTORS: Connecting With Ancestors; My Ancestors and Lineage; Indigenous Ancestors of the Land Upon Which I Live

CO-CREATING A NEW EARTH WITH NATURE AND HER SPIRITS: A New Earth; Start with Yourself; I Am Called to Energize, Harmonize, and Bless Wild Water; Ecosystem Restoration with the Fairies; Spirit Mate

The ability to communicate with nature spirits can be one of your gifts—to yourself, to your community, to the world. As Francis Weller says in *The Wild Edge of Sorrow*:

> Deep in our bones lies an intuition that we arrive here carrying a bundle of gifts to offer to the community. Over time, these gifts are meant to be seen, developed, and called into the village at times of need. To feel valued for the gifts with which we are born affirms our worth and dignity. In a sense, it is a form of spiritual employment—simply being who we are confirms our place in the village. That is one of the fundamental understandings about gifts: we can only offer them by being ourselves fully. Gifts are a consequence of authenticity; when we are being true to our natures, the gifts can emerge.

CHAPTER 15
GOING DEEPER

GO INTO NATURE AND CONNECT

Now is the time to take yourself out into nature and begin practicing the 8 Timeless Principles for communicating with nature spirits. A brief summary of what to do:

- Go into a place in nature that you love.
- Turn off your phone.
- Give your respect and send out your love.
- Find a spot that feels good to you.
- Stop, sit or lie down, relax, breathe.
- Open your heart.
- Ask for a message.
- Listen with all your senses.
- Trust your intuition when your message comes.
- Give thanks.

You can also go into nature with a trusted friend. Connect with nature spirits together, talk with each other about your experiences. Encourage each other to keep going with this practice and to be your authentic selves. Make a date for your next time in nature together. (See the Mentoring section for more specifics about this approach.)

If you want to have a different kind of nature experience, take a child into nature with you and play! Talk about your love for nature. Look for fairies together. Ask them what they know about fairies and believe what they tell you. Practice nature spirit communication principles with them. Stay open, listen, and encourage them. Have fun! Grow the next generation of nature spirit communicators.

INSIGHTS FROM INTERVIEWEES

Barbara and Aquamarine have insights to share about children and nature spirits:

> *Barbara Thomas*
>
> I thought of something that's funny…Parent says to child, "There are no fairies." Child believes in fairies. Grandmother says to child, "Not everybody believes in fairies. It's very special that you do. Be kind to your mother because she doesn't know. But maybe you can teach her. You know something that not everybody knows, so that makes you very special. The fairies need to have people to believe in them."

Aquamarine is a speech and language pathologist who works with children in a variety of ways including play therapy. She gives these insights about playing with young children:

Aquamarine

One of the great things about young kids is that there's a playful openness to them, for the most part. Even when they're having a bad day, they're still playful beings…it's almost like they know they won't be having a bad day all day long. Because I have that playfulness with the fairies, and the understanding that there are other possibilities and options, I can use these insights as a way to shift.

For example, when I'm playing with a child pretending to be a kitten climbing a tree, I may say, "I'm going to use my magic wand and cast a spell to help you get down the tree." And sometimes the child might say, "I'm not a kitten anymore. I'm now a dragon, and your wand won't work on me." So how do you "play" your way out of that? It would be easy to go into the adult part of your brain and think the "authority" way out rather than finding the play way out. Being comfortable in the fairy realm makes it easier for me to stay in that playing state and keep the play going while I'm following my bigger plan of providing the language and communication help the child needs.

The diversity of fairy realms, which are often seen as parts of play for kids, gives you so many options to bring forth. A child may or may not relate to a particular fairy, but there's a dragon, or a tree, or a mermaid, or…

DISCERNMENT AND BOUNDARIES

As a beginner in the realm of communication with nature spirits, I found that it's a good idea to go into nature to connect in a place I'm familiar with and feel comfortable in. I've been there before, I feel the beauty and love of nature there, and I imagine that I can communicate with nature spirits there. I recommend this for those of you who are starting out.

In Camilla Blossom's course Gateway to Gaia—Land Alchemy Training, I learned important discernment and boundary practices to feel safe and clear about my connections. When I go into familiar forests to connect with the trees and their spirits, I know that it feels good and right to be there (Practice 1, Discernment). However, I needed to learn the two boundary practices below to work with nature spirits on new levels, because I was going much deeper in my connection with them.

I learned that humans have free will, and that means something important in the spirit world. Because we have free will, when we respectfully and clearly speak our intentions and invitations to nature spirits, they abide by the boundaries we set. Camilla Blossom teaches that if people are completely open to any spirits, with no boundaries, unsettled or unhealthy "spirits can attach themselves to people's energy fields and cause emotional and even physical challenges."

There is so much I don't know about! It's important to listen to experienced teachers and continue to learn as you go deep with nature spirits.

The top three discernment and boundary practices for you to know:

1 DISCERNMENT—A key practice of discernment:

Take the time to recognize how you feel when you go into a particular place.

Does it feel good to be there? If it doesn't feel good, don't go there. Your body, feelings, and intuition will let you know if this is the right place for you to connect with nature spirits, so you may proceed with joy.

2 BOUNDARY—As you open deeper to these energies of nature, this statement is important to feel and say:

I connect with beneficial, healed and whole spirits, guides, ancestors, and elementals only.

These words set clear intentions and invitations. You are letting the nature spirits know you have this boundary. Since we humans have free will, spirit beings abide by this boundary.

3 BOUNDARY—Until you become more experienced with nature spirits, this boundary is extremely important:

You can honor the beings you connect with, but do not offer healing to them.

Even if you are a healer, you want to become more experienced with nature spirits before you practice your arts with them. You don't know what you could be getting into. As I practice energy healing, I knew this boundary was important for me.

An empathic human who wants to assist when she can, Camilla Blossom found herself overwhelmed at times by the number of nature spirits requesting healing from her. She teaches: Set healthy boundaries for yourself. If nature spirits ask you for healing when it's not a good time for you, you can always let them know. "I'm sorry, but I'm not available for that today."

From Camilla's book, *Sacred Spirits of Gaia*:

I highly recommend that you maintain very clear boundaries and intentions in your communication with any Indigenous Ancestors of land. This practice is for *honoring* the ancestral spirits of land. We are *not* offering healing or assistance to the ancestral spirits! There are many ancestral spirits (or soul aspects) roaming the earth that need help healing, releasing trauma, clearing their past, and crossing over. I do not recommend working with these spirits unless you know what you are doing.

Camilla Blossom has a free 24-minute YouTube video on this topic that you may want to watch if you decide to expand your connection with nature spirits: "Practices to Feel Safer and Clearer, that Open You to More Connection with Spirits of Earth." Key points from the introduction:

> This video is about how you can feel safe working with fairies, nature spirits, elementals, devas, and beings of nature. By that I mean how to set good boundaries, how to have discernment about the spirits you're working with—how to really feel good and safe about how you work with the spirits of nature, especially as you open deeper to these beautiful energies that are all around us.
>
> This video is really designed for people who are earth empaths, sensitives, lightworkers, and those who have a deep connection with the earth and the spirits of nature. So I welcome you to enjoy this way to build more powerful practices that will be helpful as you are in nature, whether it's the forest, by the ocean, the mountains, the rivers.
>
> I want to recognize that we are all One. We are like a cell in the greater whole. So I don't want to dismiss this consciousness we carry. But within that, we need to learn to be an *intact cell of integrity*. And that's why these boundaries are important.

In the video, Camilla goes through practices that are important as you prepare to deepen your connection with nature spirits:

- How do you feel?
- Ground energy
- Open to Divine/Source
- Connect with Crystalline Grid
- Clear attachments
- Make an Offering
- Ask permission (consent)
- Statement of intention and invitation
- Be clear about what you really want when you work with fairy realms.
- "Release them" after you work with nature spirits. Communicate clearly that you are done.

Be sure to take good care of yourself when you dive deep into the enchanting waters of communication with nature spirits.

MENTORING

Mentoring is sharing wisdom. A mentor can be a guide, teacher, friend or coach—someone you trust who has knowledge about what you are practicing, and is caring, compassionate, and a good listener.

In their book *Mentoring: The Tao of Giving and Receiving Wisdom*, Chungliang Al Huang and Jerry Lynch describe their beautiful mentoring process this way:

The word courage is taken from the French word coeur, which means heart. The Tao mentoring process instills courage, giving you permission to follow your heart, your passion in all that you do…Mentors can encourage mentees not to give up just when the oasis barely appears on the horizon. Encouraging others to continue the journey of heart at this point is the work of good mentoring.

This approach to the traditional idea of mentoring incorporates the Taoist teachings of self-reflection, simplicity, openness to others, and sharing of ourselves. The result is Tao mentoring, a process of learning in which the reward is not only in reaching one's goals but also in the very process of guiding and growing together. Both mentor and protégé mutually benefit from this dynamic interaction as ideas, support, and the joys of success are exchanged and shared.

Going into nature with my trusted friend Dawni Pappas is the route I began with. I lovingly call her my faery mentor. We continue to share wisdom and joy with each other after many years. Here is our mentoring story.

DAWNI PAPPAS—FOLLOW YOUR HEART

I met Dawni at a local spiritual center in the early 2000s. We started going on walks in nature together, getting to know each other in environments of beauty and wonder where we live in Santa Cruz, California.

I had been working in jobs such as information technology manager, executive assistant, scientific reports editor and secretary. At the same time, my passions have always been singing, ecological sustainability, and being in nature. I was trained as a biologist, curious about the kinds of plants and animals who live here and how they are interconnected. Although trained in science, I found myself questioning that approach. When asked what god is, I would say, "the interconnectedness of all life."

Dawni had been working in jobs such as marketing consultant, community television producer, event manager and hospitality. At the same time, many of her passions are mystical. Dawni is an elemental priestess, steeped in decades of study and practice in areas such as pagan spirituality, tarot, intuitive healing, botany, art, and multidimensional faery realms. A student of nature, she integrates her passions into a unity of knowledge and practice.

At the spiritual center, we delved inside ourselves to uncover our own light and let it shine. Dawni and I were prayer partners, learning to pray out loud and sharing our deepest depths and highest heights. We got to know each other, allowed our authentic selves to be revealed, and became comfortable being open and vulnerable with each other.

In the most natural way, we became mentors for each other in areas where we were vulnerable. I was Dawni's singing mentor and she was my faery mentor. Mentors don't try to change a person; rather, they encourage. Mentors assist in moving us to where we truly want to be, but do not yet feel capable or good enough inside. It was so important to have a shared experience of vulnerability in our friendship and our mentoring. I wrote this poem about our co-mentoring relationship:

MENTORS

Tell me about your experiences
Show me what you do
Show me how you feel
So that I may feel more deeply

We are friends
We have shared our deep selves
Without judgment
We have listened in silence together
We have cried together
We trust each other

Trust allows me to try something new
Without fear
It's just my own fear
Holding me back
Is that why the trees tell me Allow?
Allow myself to open my heart
To feel my love for the forest
For the redwoods
The bay laurels
The madrone

To feel my love for the water
The creek flowing below me
I can hear the bubbling song
Bless the water
Anoint myself with sacred water
I thank you water
I see you water
I bless you water
I love you water

Water
Open your heart and love flows
Out to the elementals
And back in return
Open
Allow

Follow your heart
Be who you are
Let your love lead the way
You are here to be fully yourself
To give your gifts
Full authentic expression
Just imagine if everyone
Let their love lead their actions
Our gifts would rain down on all

Take the risk
Open your heart
Follow the directions flowing out
Let your spirit fly

Dawni Pappas—Her message: Follow Your Heart

Working with the devas, the angels, my personal story is about following my heart. That doesn't mean I don't have fights or struggles or arguments. But I keep coming back to my heart and imagining what's possible. Because if you're stuck or you're sad, or you're in despair because of what the world looks like now—personally, globally, environmentally, spiritually—then it's time you imagine something else.

And don't be scared and discourage your creativity, imagination, joy, or love power. Allow them to come through. Follow your intuition. Follow your heart and you might just be surprised—good things will start to grow before your very eyes!

It wasn't easy for me to tell you how much I was in love with nature. We had experienced so much nature love that it probably seemed natural. But when we wanted to go out to be with people in our Tune In to Subtle

Realms of Redwoods experience, we had to do the work first. Finding that relatable, understandable, respectful, honoring place within ourselves…we had to find that with each other.

When we went into this realm together, it couldn't be just one of us knowing more than the other. It had to be co-created. Any good mentor will tell you that they are also being mentored by the mentee. So as I was more vulnerable with you about my love of nature and the call to actually take people into the woods, we had lived experience. I showed you, and shared with you, what I knew about nature and the felt-sense experience.

Then, when that felt-sense experience really happened for us together, at the same moment, in a deeper way in the forest, there was no way I could go back. I could not not-share anymore. At that sacred spot where the energy just flowed out from the water, through the air, surrounding us, you totally transformed. I was watching you and feeling it. You were practicing felt-sense. It was a beautiful thing. It was a more-than-human and otherworldly experience for me. And to share that much experience with you, I think that's a very intimate thing to do, because you're just revealed, if you're not holding back. And in that instant, we were revealed. Nature revealed their selves to us. My hands are tingling now. And so we rise in that. We had our emotions, and we had our doubts, and it brought up everything. But we were supposed to keep going forward, and that's what we did. And we did it in a sacred, healthy, timed way.

Taking people into the woods, bringing them back to nature, is mentoring. We're supposed to have shared experiences. We take people into the forest and give them time to pause. We give them a place and permission to have all their feelings. We don't ask anything from them or expect anything from them. Nothing is considered wrong or bad. We have sacred space where they can connect.

People are afraid, to a certain degree, to feel. Once we allow people to feel, they are never the same again. And then the shedding starts. Sometimes people run away from their experiences, but many times they will circle back years later, and go, "Oh yeah, I'm ready now!"

And then there's the motto on our website and experience materials: *Stay Willing—Be Wild—Be You*. There's no doubt that it was supposed to be there because it's part of the mentoring for both of us.

I do remember when that "motto" first came forth. Dawni said "Oh, that has to be in here," so I agreed. And it was interesting that we never said that to anybody verbally, but it's on the card that they take home with them. Dawni mentioning our motto while discussing mentoring is great. Because I can feel it. *Stay willing* is a huge key in what I'm doing right now, expanding my connection with nature spirits. *Be wild* is a reminder to go out in nature, feel the wildness, and let it renew my soul. *Be you* is an excellent summary phrase describing many of the messages I receive from nature spirits.

READER'S QUESTION FOR REFLECTION

How will you proceed to connect and communicate with nature spirits?

CHAPTER 16

ANCESTORS

I have been surprised to discover how often connecting with ancestors through spiritual journeys comes up as I learn about communicating with nature spirits. But the truth is that humans are an important part of nature, just like all the other species of animals. Thus, to me, the spirits of humans who have passed on—ancestors—are nature spirits.

I know many people who have had an experience of connection with a close relative after they died: a feeling, or a sign, or a message. This is often a wondrous moment that is shared with a sense of awe. I am grateful for my communication with ancestors: family members I met personally when they were alive, those I never met but heard about, and ancient ancestors completely unknown to me. Through spiritual journeys I have learned much from ancestors about where I came from, and how I am an integral part of the lineage—that unbroken chain of life since the beginning.

CONNECTING WITH ANCESTORS

I met Brenda Salgado at the Fairy and Human Relations Congress, where she was teaching about ancestors.

I interviewed Brenda and asked, "What advice might you have for people who want to connect with their ancestors, initially, and as a practice?" Brenda shared important stories that opened my eyes and heart to working with ancestors. Stories and advice from an enlightening interview with *Brenda Salgado* follow:

Start slowly, with discernment and healthy boundaries

The one thing that I like to start with, is just to tell people to be in their own discernment and intuition around this work. And to start slowly, especially if it's not something you've done before. Just like we have people in our lives who are perhaps more healthy/less healthy, more healed/less healed, the same is true of beings in the ancestor world.

I tell people that because I think that sometimes they get very scared and overwhelmed by the idea of it. Also, if you've been raised in a culture that is really reverent towards elders, that transfers to our ancestors, so we can get kind of messy around boundaries with them sometimes. So I like to share that you have sovereignty in those relationships, and you get to have healthy boundaries. It's important to know that when you start.

Ask the healthier ancestors to come forward first

As you start working with the ancestors, ask for the healthier ones to be the first ones that reach out to you: those who really want to help you on your path and purpose, and with your health and wellbeing, and that are aligned with that intention. Because these are the ancestors you can feel supported by as you begin building ancestral relationship.

As you're doing that work with them, they might turn and help you hold some of the less healthy ancestors later on, to do some work with them. But don't start there. Building relationship with ancestors is like building up a muscle. It takes repetition over time.

Consider starting an ancestor altar

I often ask people "Do you have an ancestor altar in your house? And if you don't, you might consider starting one, if that calls to you."

I knew both of my mom's parents. They lived in Nicaragua and came up to visit, especially in their older age. They both died when I was either a little girl or a teenager. But I did meet them, so that's wonderful. And then on my dad's side, I met his mother. She was here in San Francisco, so I knew and grew up with her more. She died when I was a little bit older, but still young. Those three grandparents were the grandparents I knew. When I became older and started doing medicine work and altars, I had pictures of them on my altar.

I talked more to my maternal grandmother because she's the one I always felt around me after she died. I didn't have that experience, in the same way, with the other two grandparents. Because Grandma died when I was very young, I didn't know that she was an indigenous Chorotega and was very tapped into medicines and gifts and ceremonies. I wonder sometimes how different it would have been if I'd grown up with her and she'd lived longer. I think part of Grandma's connection with me was that we shared the same kind of connection with spirit and medicine.

Catholicism and Buddhism in Brenda's life

I was raised Catholic in the San Francisco Bay Area. I was a very spiritual kid, very connected to Jesus, to Mary, to ancestors, to trees, to nature, to God. I'm grateful that when I was young, there was a really humble, wonderful priest who was the head of our school and church. He's a lot like the native elders that I know now, as an adult. He held leadership in a particular way, that I would say felt authentically humble and connected to Christ. So I'm grateful that I had that, and that I had my own kind of contemplative, direct relationship with Spirit and Creator.

Later on in life I was drawn to Buddhism, to Thich Nhat Hanh in particular. There's a particular ceremony that he did called The Five Touchings of the Earth, which I really loved. And part of it is about talking to your ancestors, asking for more guidance, and showing compassion for whatever it is they went through as human beings.

An important altar experience with a helping ancestor

One birthday I had made big altars where I had put my biological ancestors, my spiritual ancestors, my friends and family—up on these boards. And I did The Five Touchings of the Earth every day for seven days during that birthday.

That began a time when I started feeling Grandma around more. She said, "I've been waiting for you to call. We need to talk about a couple things. You come from a line of medicine people, and you don't know about that history. You need to ask your mom and one of your aunts about healings I did with their generation when they were in Nicaragua." Then Grandma said, "It's time for you to step on your medicine path. I'm going to start sending you teachers. And please check in with me when teachers come on your path. Not everyone's meant to be your teacher." She's been very clear with the yesses and nos about who I'm supposed to train with, and I really appreciate that guidance.

Whenever I would go to Native American ceremonies, often the elders or medicine people would say, "There's a little old woman standing right behind you. She's kind of small and she looks indigenous." I'd say, "It's my grandma, it's Maria, my mom's mom." And a lot of times she'd have a message for them to give to me. So she's always around, and I appreciate how present she's been that way.

I share that story because that's a story of a helping ancestor who wanted connection. It just took a little bit of intention, prayer, ceremony, and altar work to open up that relationship more deeply.

A heart-opening experience with an ancestor who appeared unhealthy

The fourth grandparent was my dad's father. But he didn't raise my dad. He didn't marry my grandma, and he left her when she was pregnant with my dad, for lots of different reasons I'm sure I don't know fully. I never met him. What was startling to me was one time when I was in my women's moon circle, and a guest was leading some journeying…and this grandparent showed up. He was very demanding and asked, "Why am I not on your altar? All the other grandparents are, and I demand to be on your altar." It felt weird.

I said to him, "First of all, thank you for showing up. Thank you for giving my father life; I wouldn't be here if you hadn't done that. And I know that you run in my blood and my bones and my veins, so I know that you're a part of me. I'm so grateful for that. And I want to be clear that it's not my fault that your picture is not on my altar. If you had raised my father, if I had met you when I was growing up, we would have so many photos of you. But as it is, I think there might be one or two photos of you from that trip my father made to you decades ago when I was in college. I don't have those photos. I don't know if Mom knows where they are. I'll ask. But I just want to be real clear that it's not my fault that your picture is not on my altar."

I was kind of shook up, as this was the first encounter I had by what felt like not a respectful tone of voice from an ancestor. I went to see a native elder not long after that for her advice. She said, "That was good, following your intuition about how to respond. You were respectful, but also clear about some things."

She looked in on my ancestor and said, "He's still doing his healing work on the other side, and there's a lot of healing work in that lineage. Tell him you do want a relationship with him, but not until it can be mutually respectful. Tell him you're going to start doing healing work on that ancestor line. Ask him to start doing his healing work on that part of the ancestral line. That we'll both start doing that work together with good intention so that we can meet some day and have a healthy relationship. And tell him, "Let's both get going on this work, and please don't contact me again until it can be a respectful, healthy relationship."

Such good advice! I said that to him, and I didn't hear from him for years. I began to do the healing work to start clearing that part of the line. Then one night on Day of the Dead, I put my three grandparents' pictures on the Day of the Dead altar. That was the next time he contacted me. He said, "Thank you for all the work you've been doing on clearing our part of the line. I'm ready to be here in a good way and be a good ancestor. You have been calling for an ancestor who held power with integrity and didn't abuse that power, and I want to help you with that…I did that."

Dad and Mom started remembering the stories about this ancestor…confirming how ethical he was, and how he never abused his power under very difficult situations in Nicaragua. Eventually Mom found two photos of him, one of which I now have on my altar. I feel like he's a really wonderful ancestor and I call him in to help when I'm doing healing sessions, when I want protection and healthy masculine energy.

In summary

In the same way that we want to have open-mindedness, curiosity, and listening with the nature spirits, we need to do that with ancestors too. Because they're complex beings just like we are. Maybe they did really beautiful things and they have beautiful gifts, but they likely also have shadow and trauma and ancestral contracts that were passed down over many generations. So just like I'm a good human being, and I also make mistakes and have trauma and healing I have to do, I know that's true of my ancestors too.

The invitation to people is to get curious about your ancestors and ask for the healthy ones to come forward. Have healthy boundaries if some show up who are still working on their stuff on the other side; don't judge them and shame them, still send them love. But you get to say, "I'd like to have a relationship with you, but not until it's a respectful one, and of good intention." And ask those healthy ancestors to help you with healing, and the less-healthy ones when they're ready.

Be patient with the process. I've been doing this work for decades, and I'm still learning things about ancestors. It might feel really huge in the beginning. Part of why it feels huge is because we've been raised in this pathologically individualistic society where we think we have to do everything by ourselves. And of course that feels overwhelming.

Many of us have also been taught not to trust our intuition, that it's our imagination and it's not real. I know I'm not doing this healing work by myself. There are many of us in the human and non-human realm doing that work together.

MY ANCESTORS AND LINEAGE

I have felt the presence of my ancestors before. But I experienced my ancestors in greater depth through guided journeys in Camilla Blossom's Gateway to Gaia course, Summer Starr's Ancestor Circle, and several Gaian Congress workshops. I discovered support and love there, and messages of encouragement in my communication with nature spirits. Here are three poems I wrote describing experiences with my ancestors.

ANCESTORS SPEAK AS OCTOBER ENDS

An ancestor song comes through me
My voice soft and shaky with my tears
The veil is thin this day
So beautiful and lacy like Hungarian table runners
Like Día de los Muertos papel picado banners
Like Tibetan prayer flags blowing in the wind for a year

I feel you, my ancestors
Mom Margie, who crossed over October 20
Dad Colin, who crossed over October 31, fourteen years before her
They are here around me now
Cheering me on in my learning about spirits of nature

Grandma Margaret is here
My Hungarian grandma who died when I was only three
Mom told me I was so much like her
We would have gotten along so well together

Grandma Margaret has come to assist me
Standing behind me, she supports me as I lean back on her
I can feel her hands on my shoulders

Light emits from her presence
"Keep going deeper," she says
"Our ancestors have danced this dance for thousands of years
We welcome you, Coleen, to the lineage"
Lovers of nature gathered together outdoors
Under the stars, fire ablaze
Praising the Divine in the flesh, in the green leaves
In the brown earth, in the ancestor stars

"Release your fears, granddaughter
Your voice is needed now
Just like it was in the time of castles, kings, knights, and peasants
This lifetime you will not be killed for your voice
Your voice in song, or your voice in writing
Now is your time
We are all here surrounding you
A spiral of women humming with you in the center
We hold you
We sing you forward
Our wisdom is accessible to you
Meet us here as often as you like
Around the fire with star ancestors above us all."

We are stardust
Dust to dust
Sacred molecules to sacred molecules

YOUNG MAN FROM THE DOUGLAS CLAN

Entering into ancestral journey
The ancient ones and the earth
Sacred space is set
To connect with the wise and well ancestors

I call upon mama incense cedar
As my support spirit for this journey
Willing to be summoned if I need her
She is protector and healer

Surrounded by forest
I ensconce myself in a hut
Made of branches with an open doorway
Call forth my wise and well ancestors
Who are interested in connecting
I sing to welcome them in

I can see fuzzy shapes of people milling about
Outside my forest sanctuary
And then one steps forth
Peeking into the doorway
Curious but a bit hesitant
A young man with reddish-brown hair
Hanging straight to his shoulders
And covering his forehead, cut as bangs

I ask if he would be my guide
Show me the land where he lives
And the vision appears
Green grassy hills with forests in the distance

I can see he is dressed in leather of animal skins
Thick panels squared off to fit
An outdoorsman
He clarifies he is from the Douglas clan of Scotland
As am I

In my journeying body at home in California
I stand, stretch my arms up
And feel roots growing out from my feet
Down into the ground
Becoming the energy of tree
To continue this journey

My guide takes me into the forest
To the massive sacred tree where his people are gathered
Honoring and celebrating together around this enormous tree
Twenty-five people hand in hand
Looking up and singing

He invites me to join the circle
I feel honored and at home
Holding hands and singing with them
I sense the energy flowing
Through the people and the sacred tree
Looking up, I see the huge branches curve downward like cedar
Pulling us together as one
We dance in love and celebration
What a delight to be with my people!

Then time for talking
I ask my guide about the disconnection
Of people from the earth we often feel in my time
Might that be relevant in his time?
I say, "You look very connected to the earth
But did something happen?"
"Invaders came
Killed many people and took our land
We are ones who survived
By going into the forest
In sacred connection with the trees"

I ask what I could do here in my time
"Bring people to nature and nature to people
Create community"
I drop to my knees
Blessing the land and the roots
Then I lie face-down on the earth
In praise and connection
I am here   I am here   I am here

Now it is time
To bring the journey to an end
I ask if I may come back and be with him again
He answers "Yes"
I am filled with love and gratitude

Young man from the Douglas clan
My ancestors of the sacred trees
This is who I am
Here and now

ANCESTORS TELL ME THIS IS MY TIME

Once upon a time, all people experienced and believed in spirits
Everything in nature was alive and everything had a spirit
Trees, animals, water, clouds, fire, soil, air
Offerings were made
Messages were received about how to live in connection
With all of nature
With all the spirits of nature

As I delve deeper into the invisible reality of nature spirits
I know I have been held in this reality before
My lineage of women healers and tree worshippers
Scrolls out in a spiral through all time

I am my ancestors' dream come true
I am here as an unbroken chain of life
from the beginning of humanity
Womb to light
Over and over again

This is true for each of us alive right now
Somewhere in your lineage
Your people knew
Your people made offerings
To the sun
To the moon
To the plants
To the animals
To the waters
To the soil
To the nature spirits

In an ancestral journey
I met with Grandma Margaret, Hungarian woman of healing
She showed me her village, mountains, and streams
Where the power of the sun was the source of ancestral strength
Gave the people life
The light, the fire, the power
Available to all

And now I recognize the fire within me
Brought forth through generations
The fire of confidence
The fire of power
The fire of destruction and creation
The fire of transformation

Ancestors tell me this is my time
To bring the knowing of the old ways of nature into full view
To alter people's way of seeing reality
Open to the invisible spirits of nature

INDIGENOUS ANCESTORS OF THE LAND UPON WHICH I LIVE

Beginning in ancient times, throughout the world, Indigenous peoples lived on and cared for land that supported their community with the necessities of life. Today, much of the current population of humans live on these same lands. Most Indigenous peoples were forced off their land in traumatic ways by invaders or colonizers who wanted their good land, and many Indigenous people died on these lands. As a result, Indigenous ancestors of the land can be present in most places that humans live today. Camilla Blossom says, "Honoring the Indigenous ancestors of lands will usher in more peace, harmony, and amends to the horrific events of the past."

During a guided ancestral journey with Camilla Blossom, we honored the Indigenous ancestors of the land we live on. I honored the Uypi tribe, Awaswas-speaking people. As soon as I spoke their name, I felt such a powerful jolt in my heart and tears flowed. I had a sense of the trauma they experienced under colonization, including the slavery and death under the rule of authority at Mission Santa Cruz.

Later I learned that among all the California missions, Mission Santa Cruz was known as the most brutal. Martin Rizzo-Martinez documents this in his 2022 book, *We Are Not Animals*:

> At the time of initial Spanish colonial settlement in 1791, around 1,400 Indigenous people from seven independent

tribes called the region known today as Santa Cruz County home. Over the next century this area became home to people from over thirty autonomous tribes from throughout the larger Bay Area, tribes such as the Uypi, Aptos, Sayanta, Chaloctaca, Tomoi, Sumus, Ausaima, Tejey, and Huocom, among others. These diverse tribes spoke one of three distinct languages—Awaswas Ohlone, Mutsun Ohlone, or Yokuts. By the time Mexican officials closed Mission Santa Cruz, in 1834, the total surviving Indigenous population numbered just over 200, a small fraction of the 2,289 Indigenous people baptized at the mission. Furthermore, many of those who died were children or infants…

…my research finds that Mission Santa Cruz did indeed become home to some of the most abusive padres. But the Indigenous community frequently challenged their authority.…Out of all the Northern California missions, Santa Cruz was the only one to face a direct attack during the Spanish colonial era and the only one where a padre was successfully assassinated. It was a site of attempted poisonings and subsequent arrests, ongoing flights of fugitives, and other examples of uprising and confrontation.

After my ancestral journey experience, I knew I had to visit the mass grave of bones from the old Mission Santa Cruz, which were moved to a cemetery not far from my house. A couple days later, I visited Holy Cross Cemetery ("Santa Cruz" means "Holy Cross" in Spanish).

CONNECTING WITH INDIGENOUS ANCESTORS AT HOLY CROSS CEMETERY
NOVEMBER 3, 2022

Today I honor the Indigenous ancestors of the land upon which I live: the Uypi, and all the Awaswas- and Mutsun-speaking Ohlone peoples. A half-mile from my home is a mass grave of the bones from the old Mission Santa Cruz, mostly the bones of Indigenous people. They were moved to Holy Cross Cemetery in 1885, with no marker, when they rebuilt Holy Cross Church. The mass grave is located at the edge of the cemetery. The rest of Holy Cross Cemetery, founded in 1876, contains many stone markers and monuments erected over graves of immigrants of Irish, German, Italian, and Spanish descent.

The mass grave contains remains of 2,583 Native Americans, Californios, and pioneers. I was present in 2016 when the dead were finally honored with three large stone markers holding the names of

Redwood trees and bronze plaques marking mass grave

the dead inscribed on bronze plaques. Most were Native Americans enslaved at Mission Santa Cruz, whose "Christian" names were handwritten in Mission records. The marker stones are surrounded by seven redwood trees planted in a sacred crescent, honoring these ancestors. Among the names in bronze are images of important relatives of the Ohlone peoples: eagle, acorn, hummingbird, butterfly, and condor.

Now I stand at the opening of the sacred crescent and speak, "I honor you, Indigenous ancestors—Uypi, Awaswas-speaking peoples. Thank you for stewarding this land for millennia before the settlers came. Thank you for showing us, in this time, that we are relatives with all beings, seen and unseen. Your descendants, the Amah Mutsun Tribal Band, continue to steward this land, as Creator has given them this sacred responsibility for all time."

I move inside the crescent of redwoods and make offerings. Slowly, I walk in a large circle, offering tobacco with great gratitude. Then I walk in a spiral toward the center, offering cornmeal. "I ask permission to be here with you today, Indigenous ancestors, honoring your gifts to this land and to our world." Feeling my hands tingling, I know the connection is made.

I begin reading the names of the dead in bronze, starting with the first group who died in 1791. These names were given by the Mission enslavers, in their "Christian" language. The names of native people are marked with a redwood tree symbol. Almost every name is of a native person.

Tears run down my face and onto the ground. An unknown chant comes forth from deep inside. I sing the chant…and continue singing. Indigenous peoples were captured, brutalized and enslaved at Mission Santa Cruz for more than fifty years, until the vast majority died. A horrendous genocide of these Indigenous ancestors took place on the land of Santa Cruz, Holy Cross, this land where I live.

I decide to read every name on the markers—all 2,583 of them—to honor these ancestors, chanting and feeling and crying. A number of the given names are also found in my current extended family: Manuel, Silvina, Monica, Rios, Natalia, Fernanda, Luisa. Other names are found in my network of friends: Barbara, Miguel, Salvador, Juan, Fernando, Reyes, Maria, Antonina, Magdalena, Cristina, Esperanza, Jovita, Paula. I feel connections through time and space. It takes me two hours to honor each individual, and I am grateful for this experience.

Names of the dead and hummingbird relative in bronze

I go to the seven redwood trees and touch their leaves. I smell their powerful scent on my hands. I touch their trunks, their bark. I hold them and know their roots are holding the bones of Indigenous ancestors below my feet. We are all connected. *Oh yes.*

I give great thanks for this day of honoring Indigenous ancestors. I sing an ancestor song in gratitude, my hands tingling, open to this land in a new way.

READER'S QUESTION FOR REFLECTION

What might you learn by connecting with ancestors?

CHAPTER 17

CO-CREATING A NEW EARTH WITH NATURE

We are in the midst of big changes. The impacts of human behaviors are being seen and felt: deforestation, pollution, systemic racism, climate chaos. At the same time, millions of people all over the world are taking action for love, healing, nature, and community.

We all need to ask ourselves: "What is mine to do?"

Each of us has unique gifts to give in service to all life, gifts that feel authentic deep inside. The great thing: **What is needed now is the full engagement of the diverse variety of our authentic human gifts.** Whatever your authentic gifts are, give them. When we give our authentic gifts in service to life, we are energized, we feel fulfilled, and all life benefits.

In the past, I have taken action with others to support the ecological, political and social causes that resonate for me. At this point in my life, my main focus is working with nature and the spiritual world and connecting people to nature more deeply, a combination that allows me to give many of my authentic gifts.

This chapter includes additional ways I am learning to work with nature to co-create a new earth, along with two examples from interviewees. The sections that follow:

- A New Earth
- Start With Yourself
- I Am Called to Energize, Harmonize, and Bless Wild Water
- Ecosystem Restoration with the Fairies
- Spirit Mate

A NEW EARTH

We can no longer go on as we have been. Let us come together as communities of people, and as people in community with nature and her spirits, to bring forth a new earth. An earth based on love and respect for all. We have a lot of healing to do, for people and for nature.

I ask the spirits of nature, "Do you have specific directions for these times to enhance the protection and restoration of the land?" They say, "Ask us specifically for land wisdom." I know the land and people are intimately related.

Nature's messages to me are most often about my own human behaviors. Nature spirits want to help me to change in ways that assist Mother Earth in the transformations that are needed.

I recognize that we, as humans, do need to change. We need to listen and act in concert with nature to co-create a new earth. Like a band of musicians bringing our gifts and creativity together, what will emerge is a new harmony and rhythm.

START WITH YOURSELF

We humans have a tendency to point our finger at other people, especially those whose greed and heartlessness destroy the land and people we love. But we need to realize and remember that when we point a finger at someone else, three of our fingers are pointing back at ourselves. It's important to recognize that each of us carries some responsibility for what's going on in this world, and we can change our attitudes and actions. Changing old habits and releasing old indoctrination can be hard, but so much is possible when we do.

Yes, we can change. Yes, we can develop our authentic gifts and give them in service to all life. Yes, we can learn new ways of being.

Interviewee Ron Hays talks about the concept of sovereignty, and how fully embracing the sovereignty of oneself and all beings is a change that can make a big difference:

> *Ron Hays*
>
> There's a word within incarnational spirituality, a common word but it's used in an uncommon way, "sovereignty." Sovereignty means that any being, whether it be a bush, a tree, or a human, has the capacity, but moreover the right, to develop and live its own being life in its own way.
>
> If we saw every being…every animal, plant…as a sovereign being, and respected that, the mutual respect that would come from that understanding and communication would shift our whole planetary existence. I think that's the key to assisting an emerging new world.
>
> When we come into our own sovereignty, we know we are not a product of the media environment around us. When we recognize our own self, our own presence, and our own sovereignty, we have the capacity to recognize the sovereignty of all the beings around us. Then we shift. And then our world shifts.

I AM CALLED TO ENERGIZE, HARMONIZE, AND BLESS WILD WATER

This practice is a beautiful way to connect with the element of water, the Indigenous ancestors, and the sacred spirits of water from the ocean, rivers, lakes, creeks, wetlands, or any wild water source. The energetic harmonizing of wild waters is one way I choose to co-create a new earth with nature. Water is life!

I have a close connection to the element of water, from the creeks in sacred forests to the larger flowing rivers, to the magnificent oceans. Learning more and going deeper with the element of water feels natural for me. Here are a few of the important experiences that led me to embrace my calling more fully.

Water is life—*Mni Wiconi*—as the Standing Rock Sioux people taught us to realize and embody in 2016, during the long protest to stop the Dakota Access Pipeline for oil from poisoning their water source, and all beings' water sources downstream. Sacred Water.

The Pacific Ocean is only a couple miles from my home. I go there often, walking barefoot in the sand and in the water coming up and down the edge of the continent. The ocean is Yemaya, Yoruban goddess and the mother of us all. In the 1990s, a visiting Yoruban elder performed a divination for me and determined that Yemaya was the orisha (goddess) with whom I am deeply connected for guidance and protection. When I walk by the ocean, I often sing to Yemaya, songs I learned in California and in Cuba. I love the 6/8 polyrhythm that accompanies my favorite song to Yemaya. Whether playing it on my conga drums at home, clapping the rhythm on the sandy beach, or using stones as my claves on a rocky beach—I sing to Yemaya.

During the wildfire here in the Santa Cruz Mountains in 2020, I would go to the ocean and ask Yemaya to use her powers to change the weather, to release fog to slow the winds fueling the hot, dry days of fire. That was *my* idea of what this powerful orisha could do...until I got the message from her: "I know what to do better than you do." I bowed to her power and presence, happy to be humble, grateful for an answer from a powerful goddess.

Creating a sacred water wheel

In the Gateway to Gaia—Land Alchemy Training course, we learned about creating a sacred water wheel in ceremony to bless and energize wild water sources for their protection and healing. My teacher Camilla Blossom traveled around the United States with her friend Marshall "Golden Eagle" Jack, member of the Paiute and Washoe Tribes of California and Nevada and Indigenous ceremonial leader and healer. The water wheel ceremony came through Marshall. Marshall and Camilla worked with people in community to bless their wild water with this ceremony. Camilla taught us how to do the ceremony, and I did mine with the ocean waters of Yemaya nearby.

The water wheel ceremony uses crystals to energetically connect wild water sources to each other through the crystalline grid. The crystalline grid can be visualized as lines of energy that crisscross the earth. In her book, Camilla quotes what crystals have told her:

> The energy of crystals is magnetic with an attracting resonance that gathers the intentions of its carrier. In addition, crystals talk to each other, communicating through the Crystalline Grid of the planet. It is in this linking that they create a greater web of strength for each individual crystal, and what it touches, as in Unity there is strength.

You, too, can learn to create a sacred water wheel to bless and energize wild water. In the Notes by Chapter section, you'll find a link to Camilla's YouTube video on this topic.

For my experiences in this realm, I go to Seabright Beach to honor Yemaya, mother of us all, and Seaweed Woman, my inner ocean goddess. I make an altar of seaweed, blue mussel shells, and my blue-and-silver Yemaya beads from Cuba, surrounding a glass jar I will use to bring ocean waters home for the water wheel ceremony.

Seabright Beach altar to honor Yemaya and Seaweed Woman

Barefoot in the ocean flow, I sing and dance in honor of Yemaya. I feel strong connection and love. I bless her and thank her for life, for the beings within her waters, for the fog and water cycle. I gather her waters in the glass jar and head home to energize and connect her waters with all earth water through the water wheel ceremony.

I create the water wheel next to my land altar. My intention is to bless and honor all the waters of earth—to protect and love all water beings and the water itself. In the center is a beautiful glass bowl, with twigs forming two interlocking triangles around it. I pour Yemaya ocean water into the bowl and place clear quartz crystals at each of the six triangle points, then rose quartz crystals next to each point. I place a rose quartz heart inside the bowl, love energy for the divine feminine, and finally add a number of tiny "worker-bee" clear quartz crystals to energize, for use in the future to bless and harmonize the wild waters in the Monterey Bay region.

Water wheel rose quartz heart and clear quartz crystals

After singing, sending love, energizing, and blessing, I point the interlocking triangle crystals outward to connect to all the earth water wheels and the crystalline grid, for the collective power of water, for protection and blessing. My whole body is tingling with this ceremony.

At completion, I remove the tiny worker-bee crystals and rose heart crystal for future blessing, then prepare to take the energized ocean water back to Yemaya.

Returning the energized ocean water

I return to Seabright Beach, returning to the ocean, Yemaya, with her water blessed, energized, and connected to water wheels and the crystalline grid of earth. This is a sacred act of love and respect. Along with my Yemaya beads, I wear my blue and white necklace of waves from Hana, Maui, connecting my heart to that sacred place and to Seaweed Woman, my inner ocean goddess.

The day is warmer than usual, with a slight breeze and clouds streaming across the sky of blue. Walking down the beach with few people present, I notice the tide has risen since I was here four hours ago, and the seaweed and shell circle I made has been taken away by the sea. I place the glass jar of energized ocean water by a giant kelp holdfast that has come ashore, take off my sandals, and walk into the water to sing and dance in honor of Yemaya.

Returning to shore, I bring the glass vessel of sacred water back with me, feet on sand with ocean foam and water washing in and out. Holding the water out and up, I praise Yemaya, Seaweed Woman, water deva, water elementals, mermaids, water dragons, undines, whales, dolphins, seals, seabirds, fish, kelp, plankton, jellies, mollusk, all life of the ocean seen and unseen, fog, the water cycle, oxygen bubbles produced by the ocean plants, carbon dioxide absorbed by the ocean. All. Blessing the balance, health, and joy of all. Returning her sacred water with its new energy connections, I pour it gently from the jar into the water moving onshore.

Aaaahhhhh…Release and let go in joy…Walking the beautiful beach as the sun moves closer to the watery horizon, I collect blue mussel shells to take home for my land altar, and to hold the energized worker-bee crystals awaiting future water blessings.

Blessing and harmonizing wild water with energized tiny quartz crystals

The worker-bee crystals begin to bless, harmonize, and energize my regional water sources in the Monterey Bay area. The tiny clear quartz crystals are natural minerals. They will do their energetic work and smooth out over time, like other stones in water. Each time I go to a body of water to offer these blessings, I feel deeply connected with Gaia and the water. Whether I go alone or with trusted friends, I feel called to this sacred work with water, giving our prayers, offerings, ritual, and song. I love singing the "Algonquin Water Song," written by Jen Rose for the Circle of All Nations Gathering in Canada. The Indigenous grandmothers there say, "Water is the lifeblood of our Mother the Earth. Water is the lifeblood of our body. We invite women everywhere to sing this 'Algonquin Water Song.'"

So far these worker-bee crystals have blessed, harmonized and energized the Pacific Ocean at Seabright Beach, Fall Creek, Arana Creek, Pinto Lake, Struve Slough, Pajaro River, Elkhorn Slough, Salinas River, Carmel River, Pacific Ocean at Asilomar, and Big Sur River.

ECOSYSTEM RESTORATION WITH THE FAIRIES
Michael "Skeeter" Pilarski Workshop

Skeeter held a workshop at the 2022 Fairy and Human Relations Congress called "Ecosystem Restoration with the Fairies." Fairy Congress was held in a beautiful valley surrounded by farmland and towns. He began with an overall description of fairy beings that I found helpful in my learning process:

> This is Fairy 101, "The Fairy Beings." All living things—all trees, insects, plants, animals, the soil microbiology—have fairies associated with them. But also the rocks, the stones, the waters, the air itself—fairy beings are associated with everything you see, including this rug and these old poles, and your body. Everything has fairies associated with it. So if we can communicate with them…
>
> Let's look at the levels of landscape fairies.
>
> 1 There are *small fairies* that are working with small-scale aspects of nature. The blackberry has fairies associated with it. Individual trees have fairies associated with particular trees. Same with the fungi that live in the space here under our mats, and individual birds, deer, squirrels.
>
> 2 Then there's a fairy being, the *deva*, that works with this whole little valley. The *deva of this valley* oversees all the smaller fairies in the valley, that work with the plants and the animals here. But within the deva of this valley is a big orb, you might imagine, that goes below and above ground. The deva's consciousness encompasses all of the little ones, and this whole space. They're in touch with the whole space, and they look after it.

3 The stream comes from outside this valley, it comes from the farmland out there. So the *watershed deva for this stream* oversees and encompasses all of its watershed, from the ridgeline all the way down to the stream. It's not just the water, it's the whole watershed. So that being encompasses, say 20 square miles, it encompasses everything, including this valley.

4 And then there is the Willamette River watershed, the bigger watershed which includes the stream watershed and this valley. The *Willamette River watershed deva* encompasses every river and stream, and all the land, in the whole watershed.

5 The Willamette is a tributary of the Columbia River. And the *Columbia River watershed deva*—now there's a magnificent, large being—and its consciousness encompasses all these different levels.

It can be called "Nesting Levels of Consciousness." The larger encompasses the next size down, and the next size down. So the largest encompasses the consciousness of everything in its realm. They are truly the local kings and queens…the old term from our feudal past. In the fairy realm, you oftentimes hear of the kings and queens of the fairies…well those are, you might say, *high devic beings* that focus on very large areas or large things. There is no single king or queen, though Pan and Anya are way up there, but there are so many levels of beings.

Humans can tap into the small level, the next up…all these different levels. It's interesting, you'd think it would be hardest to connect with the big ones and easiest to connect with the little ones. But actually I find it easier to connect with the big ones than the little ones, because they have a really expanded consciousness. They are super-smart, you might say. Whereas the smaller ones have a more limited consciousness, they are more localized divine beings, as we are all divine.

During the workshop, Skeeter also led a journey for participants to communicate with the fairy realms to ask how humans can help the land where we met. Here's a sampling of messages received by participants:

Michael "Skeeter" Pilarski

For a few minutes, I'd like us all to tune in to the deva of this particular valley where all these grand trees live. The deva that's connected to this place is really special. Here, we attract the fairies and they flock here, so they're very receptive here. This is a great place to communicate with nature and the fairy beings because they're all clued in. They're happy that we're here and are waiting for us to talk to them. So let's take a few minutes here to tune in to them. Ask if there's anything humans could do to restore this place. Are there any requests they have? How can we help this place?

After the journey, participants shared their messages:

1 I saw a face of leaves when I was connected to the valley. It was new leaf and all the phases to decayed leaf at the end. But in the middle I could see a face, a divine being, looking at me. Kind of changed to a flower in the same pattern of new to old. Felt like it's been here a long time. Showed me the whole valley. It was real pretty. I kept seeing the ivy. Looking at it, I thought "That's a big job!" I was also thinking that last year when we came, we were stripping ivy from the trees. And I wondered how that

worked out…Was there any progress? Did it help? Was it just symbolic? Did it make a difference?

Skeeter: Anybody have any further information on the ivy removal, or curbing?

2 I was walking up to the pond and I met a tree that actually wanted to express its gratitude because the ivy had been cut. So even though the old vines were very much there, the bottom three feet of vines had been taken off. The tree was thriving and doing really well. And it expressed gratitude. So from the tree spirits there's definitely an appreciation to do that work, of removing the bottom parts of the ivy so they don't continue to go up the bark.

3 I also heard about ivy. I saw this bright, shimmering golden energy being of radiance that was overall grateful for the way we treat this land. But it also mentioned the ivy, that there can be some intentional work, working with the fairy of the ivy and the fairy of the tree, to collaborate on how that process should go.

4 I also picked up in that same vein, from the deva in this valley, they were welcoming. "If you're going to be here, and you're welcome to be here, then you need to step up. If you're not going to invite the fire any longer, then you guys need to step up your game in regeneration in the forest." I think the ivy was directly related to that. And some of it was…there are places we should just stay out; this wasn't one of them. And if you're going to be here, it was a longing to be observed. That was somehow beneficial was my sense. A lot of this is so very alien to me…

Skeeter: One of the roles of humans on the planet is to love the trees, and love the planet, and love the places. So they are loved, expressed internally or verbally or with offerings or with work made visible. We are supposed to be giving love to the planet. That's part of our job, to appreciate and love and respect and observe.

5   What I got from the elementals: Be mindful how you walk on the earth. Especially being water, we are conduits to above, so they require us to communicate. To do some mindful walking, preferably barefoot, and also to be really mindful about not trashing. And if you see trash, pick it up, and they're so grateful.

6   I kept having this visualization, these images of us being in groups together, being really silent, and listening more. We do that some, but I just kept getting the sense of more being in silence together, with primacy oriented towards listening. Just amplifying the times that we are in groups…that really helps to create the portals.

7   That journey was like a way of feeling hugged from flowers, from nature, from their colors. And the way that we can bring hugs and love back to them is by visualizing colors. The colors red or orange are like a warm heart, and you can send it when you want it to be uplifting for a plant. It's hugs and colors.

8   These two majestic knights here, these big trees, are what I connected with first. But they also asked me a question, partly for me but partly for us as humans, "How do you expect to be part of our healing, if you don't heal yourself first?"

9   I saw a fairy that was in bright light, green and yellow.

I saw his or her face looking at me smiling. The thing that was spoken to me was about greed, how to address greed in people. I still need to ask more about that, but that word came up and was actually red. Telling me to look at that and see why people are so greedy. Like me, I guess. The destruction that's happening in the world right now.

10 What came up for me was love and openness. The kind of love that we're putting forth in the ground and the air is to unify the realms, and openness for that.

11 I was given the instructions to look at the tectonic plates and, in the case of a massive earthquake, the way that the rivers would change. To do the work of knowing what will happen in this valley. And to start reaching out to the people around us and start creating plans and understandings with our neighbors. And then… with our human relationships, to create this space as a sanctuary, but also to keep it safe in case of escalation of greed in city areas. To be able to protect this land by offering it as a place of love, and with love. And building those relationships now, so that in any future event, those networks are also going to be here to support the earth network.

12 I connected with the landscape deva, and my sense was that there are species missing. My first take on that was, from the little I've seen, that the lily family is probably missing somewhat—the trilliums, the Solomon's seal, the twin flowers. My sense was not to plant them, because it's not like you can easily plant those wild beings, but to invite their devas to the land. Invite them to come in.

13 What came to me is expressing gratitude through music. Through singing, through toning. That's a gift

that we can all give. It's a vibrational thing as well as a beautiful sound.

14 The message I got was about having more quiet time. Not to negate the music and celebrating, but it's the in-between noise we all make and have in our heads. To find more quiet and peace to be able to tune in and enjoy the sounds of nature. And to be able to find the answers or solutions by separating ourselves from all the busyness and stress and noise of humanity.

(Skeeter's messages) I got a number of things:

They recognized the human culture we live in and said, "Make this place a refuge. It is a refuge, but it should be designated as such. Is there some way this place could get a formal protected status so it can't be developed or destroyed?" *(participant suggestion: land trust/ conservation easement)*

They would like parts of this land to be just wild, and that we should stay out of.

I asked about invasive plants. "Should we be dealing with the invasives?" They said, "That's not a big worry. The natives here are so strong, they're not going to get pushed out by the non-natives." However, I do hear this thing about the ivy…people have been working on reducing the ivy here. A lot of the trees would be happier if the ivy didn't go up into the crown—it makes them heavier, and they could get blown over more easily. But they also said "Oh, we really like the hydrangeas over there, don't mess with that," even though they are non-native.

They also mentioned, "Have interpretive signs that really give people the idea about how special this place is, and that it should be treated with care."

> And in terms of the watershed around it that's feeding into this valley, they said, "Oh it would be really nice if you could stop people from using polluting chemicals and pesticides out there. And get more of that land into agroforestry. More of the system around it needs to be wild."

This workshop confirms that we can ask for specific advice, both about ecosystem restoration and regarding how humans can work with the fairy realms to bring forth positive change. As you can see, many messages are about changing our own human behaviors, and other messages are specific to the location in question and the surrounding region. Now it's up to the humans to take action on both of these kinds of messages.

SPIRIT MATE

Summer Starr's experience

I met Summer Starr at the 2022 Fairy and Human Relations Congress. The interview with her opened my eyes to how deep the relationships with nature spirits can be. Summer has a faerie partner, or spirit mate, whose name is Duwænem. They have a spirit marriage, as well as faerie children. I had never heard of that kind of marriage before. There's always something new to learn in the realm of nature spirits!

> To give some context about spirit marriage, in my ancestral work I have one line where the women would marry the land or waters. They were nomadic people and this was their way of getting to know the new places. I've even come to know that they had children.

Mixing between humans and the fae has been common throughout history. Some may look to the myth of Melusina for this, where there were human descendants. It is my belief that the reason she could take human form was because Melusina was also mixed. I also think this is why my faerie husband and I can connect so easily, he is a product of such a marriage…that he has human ancestors as well as land faeries and even Sidhe. So there is a lot more mixing that is going on than we humans like to think. I think that these kinds of border-crossing unions are part of nature's creativity.

Summer has a blog on her Making Kin website, and a piece called "Spirit Mate" tells the story of her relationship and marriage to Duwænem. You can read that to open your mind and increase your understanding of possibilities. She ends that piece with these paragraphs:

> My connection to nature spirits has increased since our marriage. It is like I'm a conduit for that consciousness, and sometimes when I touch a tree it feels like I'm creating a telephone connection between the land of my love and that tree.
>
> This reality may be hard for some to understand and some may call it imagination. To those I would say that the imaginal, where the fae exist, and the imagination are inextricably linked. Think of it as the imagination as the medium that receives input from the imaginal. Well, receiving is too passive…it is co-creating which honors that we are actively shaping multiple levels of reality with our hearts and minds.
>
> There are many ways for us to relate to the world around us and a spirit mate is just one. I share this important

part of my life in order to open new possibilities to others. Since my marriage I have discovered other people who have done something similar, including an ancestral guide, and this has helped to normalize this. I hope to pay that forward to anyone who also is discovering relationship in this new way.

Summer and I became classmates in the Gateway to Gaia—Land Alchemy Training course. She told me more about how deepening the relationship with her spirit mate has increased her connection with other nature spirits.

> I keep asking the question "what are we supposed to DO together," as in what is the work that my faerie partner and I are supposed to take on together…or even in the land alchemy work with Camilla, what work am I supposed to take on with these other connections with nature spirits through my land council? And I keep getting back that a big part of what is needed is to just have a heart connection with them, an openness to recognizing their existence and to communicating with them. That these interactions are affecting a kind of entrainment with the forces of nature in me…they are evolving and changing me.
> They have asked a few things of me, like placing certain stones on the land, and making a vibrational essence of the land and distributing it. And I have also seen the impact of just bringing the land council together, so the different beings are collaborating together in new ways (so in that way we are connectors), but mostly I have been getting the message, "Connect with us and it will change you."
> And that connection really has changed me. I just flew down to San Diego on Friday and on my flight down I

was connecting with some sylphs, and then when we landed I had an incredible sense of connecting with the spirit of the land and water here in San Diego. It feels really important for me to connect with the spirits of place when I travel, because otherwise I feel like I am sort of rudely dropping in on someone's home unannounced. The ease of these connections has definitely increased for me with the Norse animism work and my marriage as well as the land alchemy work…the more I consciously work on these connections, the more they happen in surprising ways. I wasn't expecting the sylph connection on my flight, but when I was meditating it sort of just called in the conscious connection.

I know that when I first married Duwænem that my connection to other nature spirits really increased. I had an experience where I laid my hand on a hawthorn tree and I could feel this language I could not understand being exchanged…kind of flowing through me between the tree spirit and my beloved. I think this is a really good reminder that relationships matter…sometimes in the human world we can get transactional about our interactions with others, but just connecting places and beings has its own importance that goes beyond some tangible outcome, but is more about the organic way of the earth where things are not so separate, they are "interbeing".

I actually had a wonderful experience of this interbeing on a recent forest bathing exploration with a group. As I sat by a rushing stream I felt how the vibration of the sound of the stream was part of the plants, the soil, the rock. The stream was not its own independent being, but the consciousness of it was part of everything that surrounded it. The rock was still its own being, too, but there was an overlap of consciousness, so the stream was in it too.

I really see this with my faerie partner. He has his own sense of being, but he is also the entire forest. There is no clear distinction. We humans have forgotten this. In fact, with our colonization and our constant migration we have forgotten our connection to land, which was really an interbeing. In fact, with the removal of indigenous people from their land we actually damaged these kinds of interbeing connections. It makes the environmental destruction we are doing a lot easier if we don't see the earth around us as part of us (or as animate in its own right).

I feel like this call back into relationship and recognizing our interbeing is a huge part of my work with the fae, and why they are always saying "just connect with us" when I ask about our work together.

I get a similar answer from the local Lummi tribal members. There are a lot of folks from western colonizer backgrounds who want to know how they can support the tribes and I often hear, "Just establish an authentic relationship with us and we will see what will evolve from there." I feel like the same is true with the nature spirits.

Summer's experiences and her depth of relationship with the land and nature spirits call forth *interbeing*—the inextricable interconnectedness among everything in the physical world and the spirit world. The false thought that we are separate from each other, from nature, and from spirit has been damaging to people and to the planet. The message that Summer received from nature spirits, "Just connect with us," is what this book seeks to facilitate. Nature spirits do want to talk with us. We can learn how to listen and understand, and then take appropriate action for the benefit of all life.

READER'S QUESTION FOR REFLECTION

Consider your unique qualities and authentic gifts. Which of these are connected to your desire to communicate with nature spirits? Which of these do you feel called to offer in service to all life?

CONCLUSION

Being in contact with nature spirits brings a new sense of excitement and wonder into your life. You'll notice your curiosity and intuition expanding as you open yourself to connection and listen for messages. When you receive messages, you'll discover what nature has to say about regenerating life—your own personal life, and the life of the natural world. Then you will realize how you can turn your messages into action.

Nature spirits want to talk with us so we can take action that is informed by nature's love. Love in action. Learning to listen to the more-than-human world empowers us to become nature spirits' partners here in the physical world. People can be a force for good, shifting our history of separation from each other and nature that has led to significant destruction of the natural world.

All of us have unique gifts that feel truly authentic. Taking action to recognize and share these gifts is the encouragement that I have received from nature spirits. I inspire people's love of nature through my work as a nature guide, my singing, and by writing and speaking about my experiences with nature spirits. This inspiration will result in people taking their own kinds of actions in untold ways—embodying the ripple effect.

I encourage you to recognize and share your authentic gifts to bring forth the powers of caring, healing, and love for people, all of nature, and nature spirits.

I look forward to hearing how you make this way of life your own. Please feel free to contact me on my website: www.coleendouglas.com

PRACTICES BY CHAPTER

Ways to engage your connections in the human and more-than-human world

CHAPTER 1 Science-Trained Mind and Nature-Loving Heart
Learn to build relationship with local Indigenous people
Learn directly from Indigenous people about caring for the land and all the life within it

CHAPTER 2 Tune In to Subtle Realms Of Redwoods
Connect with the forest in new ways, using all your senses

CHAPTER 3 What Are Nature Spirits?
Learn about the huge diversity of nature spirits
Talk with experienced nature spirit communicators

CHAPTER 4 Why Do We Need to Hear from Nature Spirits Now?
Do what we can do to encourage life
Consciously ally ourselves with nature
Nature needs us to use our free will to ask for spiritual help (Barbara Thomas)
Learn from Indigenous people to recognize all beings as sacred relatives

CHAPTER 5 Energy Healing and Receiving Messages in the Forest
Ask for a message from the trees
Stay open, continue to connect, it will become easier

CHAPTER 6 Sacred Forest Song
Use your authentic expression to assist you in opening yourself to nature spirits

CHAPTER 7 Principle 1 — Release Judgment
Recognize a knee-jerk reaction
Question a habit of resistance
Move towards love

CHAPTER 8 Principle 2 — Respect
Honoring
Ask permission
Land acknowledgment
Offerings

CHAPTER 9 Principle 3 — Love
Share love energy in nature

CHAPTER 10 Principle 4 — Ask
Ask for relationshp
Ask for assistance

CHAPTER 11 Principle 5 — Listen
Listen deeply by opening all your senses
Seamless communication (Camilla Blossom)
Find your awe, open to receive, reflect and describe (Kami McBride)

It's like a prayer, open and receive (Dawni Pappas)
Light, sound and vibratory resonance from plants
(Pam Montgomery)
Stop, quiet yourself, tune in to what's there (Ron Hays)
Stop, listen to soft voices in your heart, write what you receive
(Zing Nafzinger)

CHAPTER 12 Principle 6 — Trust Your Intuition
Whatever you're intuiting is right for you, be open-minded and patient, be in discernment (Brenda Salgado)
Look for your unique way to express what you receive, don't compare yourself with others (Kami McBride)
Intuition is part of being you and following that energy
(Dawni Pappas)

CHAPTER 13 Principle 7 — Imagine
Imagine what nature spirits look like
Call forth imagination and release our needs about how things are supposed to be (Dawni Pappas)
Pretend you can see nature spirits and ask them questions, play like kids do (Camilla Blossom)

CHAPTER 14 Principle 8 — Cultivate Relationship
Say hello
Develop routines
Be in right relationship (Brenda Salgado)
Be curious, allow life to show itself, even if it's not what you're expecting

CHAPTER 15 Going Deeper
Go into nature and connect
Take a child into nature and play
Discernment and boundaries (Camilla Blossom)
Mentoring
Follow your heart (Dawni Pappas)

CHAPTER 16 Ancestors
Connecting with ancestors (Brenda Salgado):
- Start slowly, with discernment and healthy boundaries
- Ask the healthier ancestors to come forward first
- Consider starting an ancestor altar
- Treat each other with respect, send them love

Guided ancestral journeys with experienced teachers
Honor Indigenous ancestors where you live

CHAPTER 17 Co-Creating a New Earth with Nature
Start with yourself
Give your authentic gifts in service to life
Create a sacred water wheel and bless wild waters (Camilla Blossom and Marshall "Golden Eagle" Jack)
Ask nature spirits what humans could do to restore a specific place (Michael "Skeeter" Pilarski)
"Just connect with us," say nature spirits (Summer Starr)

NOTES BY CHAPTER

CHAPTER 1
SCIENCE-TRAINED MIND AND NATURE-LOVING HEART

film by John D. Liu: Liu, John D. *Green Gold*: https://www.youtube.com/watch?v=YBLZmwlPa8A

develop Ecosystem Restoration Communities worldwide: ecosystemrestorationcommunities.org

the Camp Fire Restoration Project: Paradise, CA, Stacey Wear, Executive Director: https://www.campfirerestorationproject.org

Ali Meders-Knight…Master Traditional Ecological Practitioner: https://californiaopenlands.org Executive Director, Chico Traditional Ecological Stewardship Program https://tekchico.org

we created the document "Land Acknowledgement and Relationship-Building Protocols with Indigenous Peoples." Camp Fire Restoration Project website: https://www.campfirerestorationproject.org/resources

Mission Santa Cruz…known as the most brutal: This book documents the Indigenous people and history of Mission Santa Cruz in great detail: Rizzo-Martinez, Martin. *We Are Not Animals: Indigenous Politics of Survival, Rebellion, and Reconstitution in Nineteenth-Century California*. Lincoln: University of Nebraska Press, 2022.

this great book/audiobook: Jamail, Dahr and Stan Rushworth, Eds. *We Are the Middle of Forever: Indigenous Voices from Turtle Island on the Changing Earth*, New York: The New Press, 2022.

CHAPTER 2
TUNE IN TO SUBTLE REALMS OF REDWOODS

Airbnb experience called Tune In to Subtle Realms of Redwoods: https://www.airbnb.com/experiences/411500

CHAPTER 3
WHAT ARE NATURE SPIRITS?

John Trudell (Santee Sioux): poet, recording artist, actor, speaker, 1946-2015 www.johntrudell.com

Camilla Blossom's 2022 book: Blossom, Camilla. *Sacred Spirits of Gaia: Co-Creating a New Earth with Fairies, Elementals, Ancestors and Spirits of Land*. Nature Spirit Publishing, 2022, p.143. https://camillablossom.com/product/sacred-spirits-of-gaia-book/

Sidhe oracle cards: Spangler, David; Jeremy Berg, art. *Card Decks of the Sidhe Manual*. Douglas, MI:Lorian Press, 2011. https://lorian.org/bookstore-sidhe

Sidhe is an ancient name for the faerie races of Ireland: Spangler, David; Jeremy Berg, art. *Card Decks of the Sidhe Manual*. Douglas, MI:Lorian Press, 2011, p.1. https://lorian.org/bookstore-sidhe

excerpts from Daphne Charters' 1956 book: Charters, Daphne. *Forty Years with the Fairies: The Collected Fairy Manuscripts of Daphne Charters*, Vol.1. Compiled by Michael Pilarski. West Virginia: RJ Stewart Books, pp.11-13. https://rjstewart.net/forty-years.htm

Fairy and Human Relations Congress: https://fairycongress.com/

Gaian Congress: gaiancongress.org

CHAPTER 4
WHY DO WE NEED TO HEAR FROM NATURE SPIRITS NOW?

In current times, this is often called biomimicry: Get inspired with the Biomimicry Institute at https://biomimicry.org/

Valentin Lopez, chair of the Amah Mutsun Tribal Band: https://www.openspace.org/stories/OS-val-lopez

Amah Mutsun Tribal Band/Culture: https://amahmutsun.org/culture

Amah Mutsun Land Trust: https://www.amahmutsunlandtrust.org/

Barbara Thomas, The gnomes say, "We cannot go on any longer alone" Thomas, Barbara. *Healing Burned Woman* DVD. Full Circle Productions, 2011. "Nature" section. "This is a call to all goodhearted people to be open to Nature's intelligence and to their own intuitive knowing." https://www.amazon.com/Healing-Burned-Woman/dp/B008G0RH6G

Thomas, Barbara. "Nature Needs Us as Much as We Need Nature" Blog #114, Nov. 1, 2022. http://www.barbarathomas.info/2022/11/nature-needs-us-as-much-as-we-need-nature-2/

Michael "Skeeter" Pilarski in 2022 taught a class at the Fairy and Human Relations Congress called **Ecosystem Restoration with the Fairies.** Quotes from introduction to the class. Fairy and Human Relations Congress: https://fairycongress.com/ Global Earth Repair Foundation: https://globalearthrepairfoundation.org/

CHAPTER 5
ENERGY HEALING AND RECEIVING MESSAGES IN THE FOREST

The Living Matrix: Film about current research and practices of alternative healing: https://www.youtube.com/watch?v=tDGdakHh5Cg

Reconnective Healing: I am certified in this type of energy healing: https://www.thereconnection.com/

Forest ecologist Suzanne Simard: Simard, Suzanne. *Finding the Mother Tree: Discovering the Wisdom of the Forest.* First Ed. New York: Knopf, 2021. https://suzannesimard.com/finding-the-mother-tree-book/

Suzanne Simard TED Talk, "How Trees Talk to Each Other," June 2016. https://www.ted.com/talks/suzanne_simard_how_trees_talk_to_each_other?language=en

Bioneers Conference: Incredibly inspiring annual conference. "Bioneers: Revolution from the Heart of Nature" https://conference.bioneers.org/

CHAPTER 6
SACRED FOREST SONG

search online for "Sacred Forest Coleen Douglas" YouTube video that includes photos in the redwood forest to accompany the song. Link to download the mp3: www.coleendouglas.com

International Conference on Forest Therapy: Healing with Nature, July 7-9, 2022: Organized by University of British Columbia Department of Forestry. Conference Website with link to recordings: https://mint.forestry.ubc.ca/events/international-conference-on-forest-therapy/

"Conference Proceedings," Conference on Forest Therapy: https://mint.sites.olt.ubc.ca/files/2022/09/ICFT2022-Conference-Proceedings-Final.pdf

CHAPTER 8
PRINCIPLE 2: RESPECT

You can learn about the Indigenous people on the land where you live: Native Land Digital, http://native-land.ca Learn more about Territory (Land) Acknowledgment on their website: https://native-land.ca/resources/territory-acknowledgement/

CHAPTER 10
PRINCIPLE 4: ASK

a course called Plant Wisdom from herbalist and plant communicator Kami McBride: Online Plant Wisdom courses help people bring the healing power of herbs into their daily lives to create self-reliance and revitalize our relationship with the plant world. www.kamimcbride.com

Zing Nafzinger describes her first attempt to connect with gnomes: Nafzinger, Zing and Faerie Kin. *Faerie Wisdom to Live By.* Powerful Potential and Purpose Publishing, 2021, p.20.

CHAPTER 11
PRINCIPLE 5: LISTEN

Pam Montgomery wrote the book: Montgomery, Pam. *Plant Spirit Healing: A Guide to Working with Plant Consciousness.* Bear & Company, 2008. https://www.wakeuptonature.com/learning/books/

Zing translated communications from a faerie named Celerine:

Nafzinger, Zing and the Faerie Kin. *Faerie Wisdom to Live By.* Powerful Potential and Purpose Publishing, 2021, pp.65-67.

CHAPTER 13
PRINCIPLE 7: IMAGINE

Then we did a "gnome dance" that Camilla demonstrated: A clear description in Blossom, Camilla. *Sacred Spirits of Gaia: Co-Creating a New Earth with Fairies, Elementals, Ancestors and Spirits of Land.* Nature Spirit Publishing, 2022, p.228. https://camillablossom.com/product/sacred-spirits-of-gaia-book/

Mariel describes what imagination is for the Sidhe: Spangler, David. *Engaging with the Sidhe: Conversations Continued*, Douglas, MI:Lorian Press, pp.11-13, 16-17. https://lorian.org/bookstore-sidhe

CHAPTER 14
PRINCIPLE 8: CULTIVATE RELATIONSHIP

My Council of Gnomes told me, "Open your heart, Barbara." Barbara Thomas's communications with nature spirits on her blog: http://www.barbarathomas.info/blog/

Brenda Salgado is a first-generation Nicaraguan-American: Find out more about Brenda and her offerings at https://www.nepantlaconsulting.com/

I select an oracle card from the Card Deck of the Sidhe: Spangler, David; Jeremy Berg, art. *Card Decks of the Sidhe Manual.* Douglas, MI:Lorian Press. "The Living Stone" p.216, "Anwa" p.210. https://lorian.org/bookstore-sidhe

PART THREE: DEVELOP YOUR ABILITY TO COMMUNICATE

As Francis Weller says: Weller, Francis. *The Wild Edge of Sorrow:*

Rituals of Renewal and the Sacred Work of Grief. Berkeley, CA: North Atlantic Books, 2015. https://www.francisweller.net/books.html

CHAPTER 15
GOING DEEPER

Al Huang and Jerry Lynch describe their beautiful mentoring process: Quote from front flap of hardback book cover. Huang, Chungliang Al and Jerry Lynch. *Mentoring: The Tao of Giving and Receiving Wisdom.* First Ed. SanFrancisco:HarperOne, 1995.

I learned important discernment and boundary practices: Camilla Blossom's course, Gateway to Gaia—Land Alchemy Training. https://camillablossom.com/gateway-to-gaia-land-alchemy-training/

In Camilla's book Sacred Spirits of Gaia: Blossom, Camilla. *Sacred Spirits of Gaia: Co-Creating a New Earth with Fairies, Elementals, Ancestors and Spirits of Land.* Nature Spirit Publishing, 2022, p.75. In the section Co-Creating with Indigenous Ancestors. https://camillablossom.com/product/sacred-spirits-of-gaia-book/

Camilla Blossom has a free…YouTube video on this topic: "Practices to Feel Safer and Clearer, that Open You to More Connection with Spirits of Earth." https://www.youtube.com/watch?v=1lQ9uuPdHS8

CHAPTER 16
ANCESTORS

Brenda Salgado is a first-generation Nicaraguan-American: More about Brenda and her offerings at https://www.nepantlaconsulting.com/

The Five Touchings of the Earth: Thich Nhat Hanh ceremony. https://plumvillage.org/key-practice-texts/the-five-earth-touchings

Summer Starr's Ancestor Circle: https://www.makingkin.com/ancestral.html

Camilla Blossom says, "Honoring the Indigenous ancestors…" Blossom, Camilla. *Sacred Spirits of Gaia: Co-Creating a New Earth with Fairies, Elementals, Ancestors and Spirits of Land.* Nature Spirit Publishing, 2022, p.73. https://camillablossom.com/product/sacred-spirits-of-gaia-book/

Martin Rizzo-Martinez documents this in his 2022 book: Rizzo-Martinez, Martin (2022), *We Are Not Animals: Indigenous Politics of Survival, Rebellion, and Reconstitution in Nineteenth-Century California.* University of Nebraska Press, pp.11-13.

CHAPTER 17
CO-CREATING A NEW EARTH WITH NATURE

…it's used in an uncommon way, "sovereignty…" Lorian is a spiritual education community. "We believe that you have a unique purpose and call that can transform the world. Learning to shape this transformation lies at the heart of Lorian's teachings." https://lorian.org/power-of-incarn

creating a sacred water wheel to energize the wild water: Camilla Blossom offers a free how-to video on her YouTube channel: Honor Mother Earth's Waters - Water Wheel - Crystals & Sacred Geometry https://www.youtube.com/watch?v=1BpqbDZCPxI Also see "Crystals" chapter of her book: Blossom, Camilla. *Sacred Spirits of Gaia: Co-Creating a New Earth with Fairies, Elementals, Ancestors and Spirits of Land.* Nature Spirit Publishing, 2022, pp.231-236. https://camillablossom.com/product/sacred-spirits-of-gaia-book/

-I love singing the "Algonquin Water Song." Written by Jen Rose for the Circle of All Nations Gathering in Canada, 2002, this song is meant to be sung by all women who feel called to sing it. https://www.singthewatersong.com/songlyrics "...the Algonquin Water Song expresses loving gratitude for the water and raises the consciousness and connection of women with Mother Nature's greatest gift. The song is easy to learn, and our vision is that millions of women will sing it, raising their own connection and awareness of the water they interact with daily even in the shower or at the sink."

"Spirit Mate" tells the story of her relationship and marriage to Duwænem: Summer Starr's blog on her website has one piece called "Spirit Mate." https://www.makingkin.com/blog/spirit-mate

Since my marriage I have discovered other people who have done something similar: Learn more in this comprehensive interview on YouTube: "Spirit Sex and Marriage with Megan Rose" https://www.youtube.com/watch?v=XqL0hzxU7nM

FURTHER LEARNING RESOURCES

Sharing my personal recommendations for further learning about nature spirits and communicating with them, and for learning more about Indigenous wisdom. There are many more venues for further learning. It's a lifelong joy!

EVENTS
Fairy and Human Relations Congress fairycongress.com
Online Gaian Congress gaiancongress.org

Tune In to Subtle Realms of Redwoods
with Coleen Douglas & Dawni Pappas
www.airbnb.com/experiences/411500

Participants say:

- I've never felt more connected to nature and myself.
- I came away energized with the feeling that I can develop a personal relationship with nature in a new way that supports my activist work for the planet.
- My 8-year-old daughter said, "This is something I'm going to remember as an adult!"

Private Redwood Experiences for Organizations: team building, retreats, and more. Contact Coleen at www.coleendouglas.com

FURTHER LEARNING RESOURCES

Global Earth Repair summits and website
globalearthrepairfoundation.org
Michael "Skeeter" Pilarski, organizer. Results of 2019 and 2022 Global Earth Repair international events available on the website. The 2022 Nature Communication and Indigenous Leadership tracks include video recordings of each session.

TEACHERS

Camilla Blossom
10-month course: Land Alchemy Training—Temple of Earth Magic
camillablossom.com/gateway-to-gaia-land-alchemy-training
Having taken this course myself, I can tell you it is fantastic! I learned about and experienced many nature spirits, how to begin communicating and co-creating with them, and how to connect with the land and waters in new ways.

Camilla Blossom's YouTube channel has many free videos "to open, be inspired, and learn how to tap into the spiritual, energetic, and magical realms of Nature—seen and unseen." www.youtube.com/@CamillaBlossom1111

Kami McBride
Online course, "Plant Wisdom: Activate Your Plant Communication Skills"
This course has everything inside to help you discover how to listen to, accurately interpret, and confidently apply messages you receive directly from the plants. This online course is available at any time, and the class materials are available to you for a lifetime.

This is an excellent course that I have taken. Kami is an herbalist with decades of knowledge about herbs and all kinds of plants.

https://www.plantwisdom.online/start-here Kami also has free posts on her blog about plant communication: https://kamimcbride.com/?s=plant+communication

Dawni Pappas
Meet Dawni Pappas, a weaver of life-affirming connections between the seen and unseen realms. As a lifelong student of Nature, she has evolved into an Initiated Elemental Priestess, Nature Guide, and Faery Mentor (including being Coleen's Faery Mentor). Dawni offers diverse experiences and healing practices, each infused with her deep understanding and reverence for Nature and the Subtle Realms. She is dedicated to bringing people back to Nature and Nature back to people. Explore her offerings and embark on a journey of connection, discovery and co-creation at dawnipappas.com and naturesembrace.me

Brenda Salgado
Brenda is Founder of Nepantla Consulting. She has taught, facilitated, and led ceremony at numerous gatherings and organizations. She is an author, speaker, indigenous wisdom keeper and healer. Brenda consults with people regarding energetic, personal, and ancestral healing. She has received training in traditional medicine in Purépecha, Xochimilco, and Toltec lineages. Brenda draws on the healing powers of the natural and spirit world to guide her work. www.nepantlaconsulting.com Contact: brendasalgadoasst@gmail.com

ORACLE CARDS
Card Deck of the Sidhe (deck and manual/book), Spangler, David; Jeremy Berg, art. *Card Decks of the Sidhe Manual*. Douglas, MI: Lorian Press, 2011. Order here: https://lorian.org/bookstore-sidhe

FURTHER LEARNING RESOURCES

Sacred Spirits of Gaia: Oracle Deck. Camilla Blossom oracle cards. Order here: https://camillablossom.com/product/sacred-spirits-gaia/

ABCs of Knowing U R A Majikal Being, Dawni Pappas, 2021.
This book is an oracle (helpful answer book) that you simply flip open to any page and receive an answer. These Majikal messages can bring you joy, create new space for your Majikal capabilities to grow and help you engage with your own circumstances in new and lighthearted ways.
Order on amazon.com
https://www.amazon.com/ABCs-Knowing-URA-Majikal-Being-ebook/dp/B09NLHWVPW?ref_=ast_author_mpb

LEARNING MORE ABOUT INDIGENOUS WISDOM

At this time, there are many ways to learn from Indigenous people: books, audiobooks, webinars, websites, Traditional Ecological Knowledge (TEK) workshops, podcasts, and more. A recent online search for "Indigenous Wisdom" generated 44 million results. Please go into your process by learning directly from Indigenous people. Recommendations to get you started follow.

Events
Bioneers Conference/Indigenous Forum
https://bioneers.org/indigenous-forum/
The annual Bioneers Conference began the Native-led Indigenous Forum in 2008. It was designed as a sovereign space for Indigenous People to bring their vision and message to Native and non-Native allies and to connect. It has grown dramatically over the past 10 years and had more than 300 Indigenous attendees belonging to 94 distinct tribes in 2018. It remains the only gathering of its kind,

bringing together Indigenous activists, scientists, elders, youth, culture-bearers and scholars to share their knowledge and frontline solutions in dialogue with a dynamic, multicultural audience.

Books/Audiobooks
Jamail, Dahr and Stan Rushworth, Eds. *We Are the Middle of Forever: Indigenous Voices From Turtle Island on the Changing Earth*, New York: The New Press, 2022.
Interviews with people from different North American Indigenous cultures and communities, generations, and geographic regions who share their knowledge and experience, their questions, their observations, and their dreams of maintaining the best relationships possible to all of life.

Kimmerer, Robin Wall. *Braiding Sweetgrass: Indigenous Wisdom, Scientific Knowledge and the Teachings of Plants*. Minneapolis, MN:Milkweed Editions, 2015.
Drawing on her life as an Indigenous scientist, and as a woman, Kimmerer shows how other living beings—plants and animals—offer us gifts and lessons, even if we've forgotten how to hear their voices. The awakening of ecological consciousness requires the acknowledgment and celebration of our reciprocal relationship with the rest of the living world. For only when we can hear the languages of other beings will we be capable of understanding the generosity of the earth, and learn to give our own gifts in return.

Online Resources
Bioneers Indigeneity Program
https://bioneers.org/indigeneity-program/
Indigeneity is a Native-led program within Bioneers that promotes

Indigenous approaches to solve the earth's most pressing environmental and social issues. They produce the Indigenous Forum, original media, educational curricula, and catalytic initiatives to support the leadership and rights of First Peoples, while weaving networks, partnerships and alliances among Native and non-Native allies. Bioneers is an innovative nonprofit organization that highlights breakthrough solutions for restoring people and planet.

Cultural Competence Circle
Based on extensive notes from our Cultural Competence Circle meetings, we created the 2021 document "Land Acknowledgement and Relationship-Building Protocols with Indigenous Peoples at Ecosystem Restoration Communities California." You can find this document on the Camp Fire Restoration Project website https://www.campfirerestorationproject.org/resources

Our final offering in February 2021 was a day-long series of webinars: "Listening to the Land Together: Weaving Connection Among Indigenous Wisdom Keepers and Ecosystem Restoration Allies."

Two free 90-minute recordings are available:

1 "Listening to the Land Together: Panel on Respectful Collaborations" hosted by Gregg Castro, February 2021, Ecosystem Restoration Communities https://www.youtube.com/watch?v=I-J4EJ1U9KA Cultural Competence Circle leader Gregg Castro (t'rowt'raahl Salinan/rumsien-ramaytush Ohlone) hosted panelists Janet Eidesness (MA) and Shannon Tushingham (PhD) who discussed their own work approaching and engaging Indigenous communities.

2 "Listening to the Land Together," Keynote Address by Dr. Cutcha Risling Baldy, February 2021, Ecosystem Restoration Communities https://www.youtube.com/watch?v=zE8wgwW0BLY
Dr. Cutcha Risling Baldy (Hupa, Yurok, Karuk) is Chair of the Native American Studies Department at Humboldt Cal Poly University and a leader in the #landback movement.

Native Land Digital
You can find out about the Indigenous people on the land where you live on this website: http://native-land.ca/
Native Land Digital is a Canadian nonprofit organization and has been creating and maintaining this website since 2015. As they say, "It is not perfect," but it gives a growing body of detailed information, especially in North and South America and Australia (with development occurring all the time in different parts of the world). Their Territory Acknowledgment guide gives good ideas about how and why to create a Land Acknowledgment. https://native-land.ca/resources/territory-acknowledgement/

Native Land Digital strives to create and foster conversations about the history of colonialism, Indigenous ways of knowing, and settler-Indigenous relations, through educational resources such as our map and Territory Acknowledgement Guide…We aim to improve the relationship of people, Indigenous and non-Indigenous, with the land around them and with the real history and sacredness of that land. This involves acknowledging and righting the wrongs of history, and also involves a personal journey through the importance of connecting with the earth, its creatures, and its teachings…Native Land Digital has a wide range of impacts on individuals, classrooms, and others who view and interact with our map and associated content. We provide a Teacher's Guide, Territory Acknowledgement Generator, a blog, and other content in an effort to widen this impact along the lines of our missions. https://native-land.ca/about/why-it-matters/

Hands-On Training Program
Chico Traditional Ecological Knowledge Training Program, Indigenous Led Land Management for Community Resilience and Shared Prosperity, Chico, California, https://tekchico.org/

MEET THE INTERVIEWEES

AQUAMARINE

Aquamarine is a speech-language pathologist (SLP) who works with children, mostly in a school setting. Play is a great way to get children to participate in speech therapy and an excellent way for adults to get into a creative headspace.

BARBARA THOMAS

Barbara Thomas is 96 years old and an active artist, writer, international traveler, and leader within an interdenominational Christian organization. She has traveled to 53 different countries, meeting her Guardian Angel at age 21, when bicycling through Europe the first year it opened after WWII. Barbara lives in the heart of a redwood forest and interacts daily with the Spirits of the Land. Angels guide her paintings, opening portals of color and shape to allow the Spiritual Dimension of Life to enter. When Gnomes started appearing in her paintings, she met Mano, a Master Gnome who became her teacher. Mano invited Barbara to collaborate with him to create his blog, Council of Gnomes; 128 blog posts have been published. www.barbarathomas.info/blog Barbara has written four books and created a four-part DVD named *Healing Burned Woman*, each telling different aspects of her life with Angels, Elementals and Mother Earth. Visit her website: www.barbarathomas.info

BRENDA SALGADO

Brenda is Founder of Nepantla Consulting. She has taught, facilitated, and led ceremony at numerous gatherings and organizations. She is an author, speaker, indigenous wisdom keeper and healer. She has over 25 years of experience in spiritual development, non-profit management, transformative movement building, and social justice. She is the author of *Real World Mindfulness for Beginners*. Brenda has served as the Director of the East Bay Meditation Center, Associate Director at Wisdom & Money, and a Senior Fellow at Movement Strategy Center. She holds degrees in biology, psychology, and animal behavior, and has received training in traditional medicine in Purépecha, Xochimilco, and Toltec lineages. Brenda draws on the healing powers of the natural and spirit world to guide her work. She is committed to co-creating a society filled with wholeness and beauty. www.nepantlaconsulting.com bit.ly/RealWorldMindfulness Racial Healing Initiative — Retreat Center Collaboration: https://www.retreatcentercollaboration.org/rhi Contact: brendasalgadoasst@gmail.com

CAMILLA BLOSSOM

"We are not here to heal the earth...We are here to heal our relationship with Mother Earth." Camilla is a mentor, land alchemist, intuitive energy healer, ceremonialist, and medicine woman. Her passion is to co-create with the spirits of plants and trees, land and water, The Grandmothers, and elemental fairy realms of Nature. She offers earth-based wisdom teachings through her Patreon community, private mentorship, land alchemy, events, and trips. Her Earth Magic offerings support Earth Guardians, Tree Keepers, and Land Alchemists in stepping into their mastery. Her most recent book is *Sacred Spirits of Gaia: Co-Creating a New Earth with*

Fairies, Elementals, Ancestors, and Spirits of Land and companion oracle deck Sacred Spirits of Gaia. CamillaBlossom.com https://camillablossom.com Patreon: www.patreon.com/camillablossom

DAWNI PAPPAS

Dawni Pappas weaves together a life-affirming tapestry of connections and gifts with the seen and unseen worlds. As a lifelong student of Nature, her journey has led her to become an Initiated Elemental Priestess and a Faery Mentor. One way she co-creates with the Subtle Realms is by taking people, along with Coleen Douglas, into the Redwoods for experiences, in California and beyond! www.airbnb.com/experiences/411500

Dawni is an organizer of the Gaian Congress and the Fairy and Human Relations Congress. She has a background in divination and spirituality. She practices more-than-human communications and is a Nature guide, author, educator, instructor, energy healer, spiritual practitioner and coordinator of Divine Events. Dawni has been co-creating in beauty and helping others to heal and remember their fullness in and with Nature for over 25 years.

For more information and consultations, contact her through: dawnipappas.com and naturesembrace.me

JAZZ

Hello World! I'm Jazz, no relation. I grew up free-range gallivanting between Appalachia and her neighboring valley and grew up with a deep love of the land. Running through fields and swimming in streams. Walking home by the light of the moon. Feeling the deep presence and spirit of the land through the shift of her seasons. I brought that love, passion, and inspiration into my adult life and use it in my art. Trying to give it feeling and soul. It is my hope that I can one day be established enough to give back to the lands that I love. https://instagram.com/blackbear_elk?igshid=ZDdkNTZiNTM=

KAMI MCBRIDE

Kami McBride is the author of *The Herbal Kitchen*. She has taught herbal medicine at University of California, San Francisco School of Nursing. Her online courses help people to bring the healing power of herbs into their daily lives to create self-reliance and revitalize our relationship with the plant world. Over the past 35 years Kami has helped thousands of families learn to use herbs and healing plants for their self-care, to prevent illness and to deeply connect with the earth and her abundant medicine. www.kamimcbride.com

LINDSEY SWOPE

Lindsey Swope has always felt a strong connection to the Earth. She earned degrees in both Geology and Geography before recognizing her spiritual call to work with the subtle realms of Fairies, Devas, and Nature Spirits. She has been involved with the Fairy and Human Relations Congress since its inception in 2001. www.fairycongress.com The Fairy Congress, as it is commonly known, started as an annual event held in the Pacific Northwest. In 2021 we were urged by our allies to start an online version called the Gaian Congress, in recognition of the expansion of our network of allies. www.gaiancongress.org There is now also an ongoing, online community platform to offer continued support for inter-dimensional relations. Lindsey's current job title is the Queen of Logistics, which involves working with delegates on both sides of the veil to create and continue harmonious relations between the realms.

MICHAEL "SKEETER" PILARSKI

Michael "Skeeter" Pilarski was one of the founders of the Fairy and Human Relations Congress in 2001: fairycongress.com. He is a fairy librarian: thefairylibrary.org. He is a life-long student of plants and earth repair. Michael has been working on ecosystem restoration for decades, and is the convener of the Global Earth Repair Conference globalearthrepairfoundation.org. Michael founded Friends of the Trees Society in 1978, and since 1988 he has taught 36 permaculture design courses in the US and abroad. He has hands-on experience with over 1,000 species of plants. Friends of the Trees Botanicals website provides education, events, and retail/wholesale sales of their organic herbs and plants friendsofthetreesbotanicals.com. Michael is a prolific gathering organizer and likes group singing.

PAM MONTGOMERY

Pam Montgomery is an herbalist, author, international teacher, Earth elder and new-paradigm thinker who has passionately embraced her role as a spokesperson for the green beings and has been investigating plants and their intelligent spiritual nature for more than three decades. She is the author of two books, one of which is the highly acclaimed *Plant Spirit Healing: A Guide to Working with Plant Consciousness* and is currently writing her latest and third book, *Co-Creative Partnership with Nature*. She teaches internationally and virtually on plant spirit healing, spiritual ecology and co-creative partnership with Nature. She is a founding member of United Plant Savers and more recently the Organization of Nature Evolutionaries or ONE. www.wakeuptonature.com

MEET THE INTERVIEWEES

RON HAYS

Ron has a lifelong interest in spiritual disciplines, spending some years in a spiritual group founded by one of G.I. Gurdjieff's students. Afterward, he learned shamanic practices with Michael Harner. Currently, he is a practitioner of the Lorian Association's Incarnational Spirituality lorian.org. He is an ordained Lorian priest and faculty member, teaching nature classes. Approximately 15 years ago, he was approached by members of the Sidhe—the incarnate, nonphysical cousins of humanity. He works closely with them, and his series of outdoor sculptures can be seen at elvengates.com. He also shares his and others' Sidhe experiences on portalsconnect.com.

SUMMER STARR

Summer Starr is a spirit-worker who helps guide people in exploring the ancestral realms, the psyche and their connections with nature. Diagnosed with bipolar disorder, she holds wisdom about the intersection of working with the unseen realms and altered mental states. Learn more: www.makingkin.com

ZING NAFZINGER

Zing Nafzinger is a creator with a master's degree in the Psychology of Therapy and Counseling, partnering with the energies of the Angelic and Faerie Realms as a counselor, teacher, writer, Reiki healer, and artist, and exploring her interests in beach mandalas, faerie energy work, permaculture, creativity workshops, and soul retreats around the world. Her books: *Mediumship as a Way of Being* and *Faerie Wisdom to Live By*. View her Faerie Meditation videos on YouTube at 'Zing Nafzinger". Follow her ongoing adventures on Facebook at "Zing Nafzinger". Email questions and joyous invitations to i.m.zing@comcast.net Phiddly-dee, phiddly-dae!

ABOUT THE AUTHOR

COLEEN DOUGLAS is a nature lover who has been learning how to communicate with nature spirits over the last ten years. She is a nature guide, spiritual practitioner and energy healer in Santa Cruz, California who recognizes the interconnectedness of all life, seen and unseen. A singer and song leader, Coleen uses her voice to touch human hearts and connect with nature spirits.

With her friend Dawni Pappas, Coleen takes individuals and organizations into the forest to Tune In to Subtle Realms of Redwoods and experience a deep connection to nature. Coleen has a BA in biology and an MA in organizational development. Her science training has both strengthened and occasionally hindered her unconventional learning process to communicate with nature spirits.

Coleen is grateful to continue learning about Indigenous wisdom and the practices of ecosystem restoration—both are needed to regenerate the natural world, in concert with nature spirits. She is a grandmother working for the benefit of future generations. Coleen believes that love is the open doorway to connection, with nature spirits and with people. Love is the power that unifies. Reconnecting people to each other, and to nature and her spirits, is the way forward.

Contact, Info, Music
www.coleendouglas.com

Nature Walks: Tune In to Subtle Realms of Redwoods
www.airbnb.com/experiences/411500

Songs
www.youtube.com/@CD621

Made in the USA
Middletown, DE
15 June 2024

55600199R00115